MODIFY IMPROVE & UPGRADE
YOUR KIT CAR

From the publishers of Complete Kit Car magazine

First published 2016
Copyright Performance Publishing Ltd

All rights reserved. No part of this publication may be reproduced or transmitted in any form or by any means, electronic or manual, including photocopying, recording or by any information or retrieval system, without prior permission in writing from the publisher.

ISBN 978-0-9576450-1-1

Authors: John Dickens, James Horsley, Ed Morton, Peter Rosenthal, Martin Scott, Ian Stent
Editor: Adam Wilkins
Designer: Sarah Scrimshaw

Printed by: The Manson Group Limited, Hertfordshire AL3 6PZ

Disclaimer
While every effort has been taken to ensure the accuracy of the information given in this book, no liability can be accepted by the authors, publishers or distributor for any loss, damage or injury caused by errors in, or omissions from, or misuse of the information given.

Publisher: Performance Publishing Ltd
Unit 3 Site 4 Alma Park Road,
Alma Park Industrial Estate,
Grantham, Lincolnshire NG31 9SE, Great Britain

Modify, Improve & Upgrade Your Kit Car

Introduction

"They're never really finished" is a common assertion of the kit car owner. Sure, there are milestones in any build: first engine start, first drive, IVA approval, registration. But the urge to personalise and upgrade a kit car never goes away, no matter whether you built the car yourself or bought it second-hand.

A significant part of any issue of Complete Kit Car magazine is given over to exactly that topic, and Modify, Improve & Upgrade Your Kit Car is a compilation of the hands-on, how-to features from the last few years. This book doesn't contain every technical feature from recent past – there wouldn't have been space. Instead, we've selected the how-to features that you can follow in your own garage, rather than anything that's overly theoretical or relates to a specific project car build. If you're looking for an excuse to get in the garage and modify, improve or upgrade your kit car, let these pages provide the inspiration...

Featuring...

JOHN DICKENS
Complete Kit Car Technical Editor, once built his own GRP monocoque for a GTM Coupé. Has built many kit cars.

JAMES HORSLEY
Has a thing for VW based kit cars. Restored an immaculate Nova and now has one with Subaru power.

ED MORTON
Built an MGF based version of the classic 1970s Nova kit car, and then did the same for a beach buggy.

PETER ROSENTHAL
Built a Westfield, owned a couple of Sylvas and now has a Garder Douglas T70. Doesn't do things by halves.

MARTIN SCOTT
Ex-Lotus man who owns several kit cars, most of which see very regular use (and one that's overdue restoration).

SUBSCRIBE AND SAVE!
www.completekitcar.co.uk | 01476 978843

Modify, Improve & Upgrade Your Kit Car

Contents

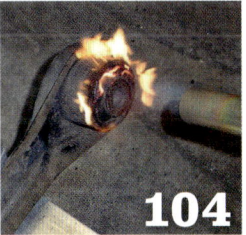

008 IN THE ENGINE BAY
Covering cooling, the fuel system, hose clips, and fitting motorcycle throttle bodies.

025 BRAKES
Everything you need to consider when speccing your kit car's brakes, plus how to bleed them with a vacuum.

034 BODYWORK & GLASS
How to make and repair GRP panels, plus fitting an aerocatch and vinyl wrapping. Also how to etch glass.

050 RUST-PROOFING
Powdercoating, plating, Waxoyling and other methods to stop your ferrous components corroding.

064 FABRICATION
How to make brackets and fit control cables.

069 ELECTRICS
General kit car wiring advice, and specifics on making intruments work accurately and upgrading headlights.

082 IN THE GARAGE
Improve your garage flooring, and find out how to correctly use impact screwdrivers and sprayguns.

095 TRIM
How to trim an interior panel, and your general kit car trimming options.

104 RUNNING GEAR
All you need to know about wheel offsets, polyurathene bushes and CV joints.

120 ROLLCAGES
What to consider when speccing and fitting a rollcage.

125 USEFUL CONTACTS
A handy directory of contacts.

Keeping Cool

Ed Morton considers the various components within a typical automotive cooling system, explaining how they work and what you may need to consider for your own car.

In my day job as a vet, I get to experience a few choice odours – but nothing fills me with as much dread as the scent of hot coolant. Being surrounded by billowing clouds of steam only adds to the comedy value of a breakdown if it's not your car. I used to own a Reliant Scimitar and a Fiat X1/9 – so I've been there.

Assuming the basic cooling system design is sound (which in some cars might be a pretty big assumption), cooling system faults such as gunged up water passages, stuck thermostats and leaks are usually relatively predictable and straightforward to rule out, although I was once stumped for a very long time by a water pump with a loose impeller. But overheating problems can sometimes be caused by faults elsewhere, such as binding brakes, retarded ignition timing or cylinder head gasket failure. Just occasionally, the temperature sender or gauge is broken. However, cooling system fault diagnosis is widely covered in workshop manuals. What isn't as easy to find out is how to optimise the cooling system in the first place, so that's what this article will attempt to do, with help from Matt Foreman at Car Builder Solutions.

RADIATOR

On the face of it, a donor vehicle radiator is a logical choice. These are usually aluminium with plastic tanks at either end, which makes them helpfully lightweight and often quite inexpensive. However, they may not be a suitable shape, and have been designed

alongside optimised fan assistance and ducting, which might not be the case in their new home.

Generic kit car aluminium radiators are often based on an early VW Polo design, as used extensively in many Seven replicas. It is very compact, at around 310mm by 460mm and with several mounting options, and manufacturers can often be flexible about the position and sizes of coolant inlets and outlets. Its cooling capacity can be increased substantially by using two or three cores, which increases the radiator's depth but not the frontal area. A three-core version will cool a 200bhp V8 effectively.

It may be more straightforward to use two smaller radiators (maybe on each side in front of the rear wheels) as a supplement to or instead of one larger one... its the solution for a number of mid-engined cars, most obviously Countach and Diabo replicas.

As a rule of thumb, the radiator air inlet should be around a third of the radiator's core area. Some control of the inlet air is essential – a radiator core resists air flow, so if an alternative route around the radiator exists, a substantial amount of the incoming air will bypass it. Good ducting can reduce the area of radiator core required by 50 percent. Exit ducting to a low-pressure area, or at least providing an easy route out behind the radiator is equally important – even though it can be very tempting to fill this chunk of space with batteries, fuel tanks, luggage space... and any other gubbins that won't fit anywhere else.

Oh dear! Every kit car owner's nightmare. But with a bit of care and thought you can hopefully avoid this.

An aftermarket radiator is a good upgrade over the usual Volkswagen Polo unit.

Modify, Improve & Upgrade Your Kit Car

In the engine bay

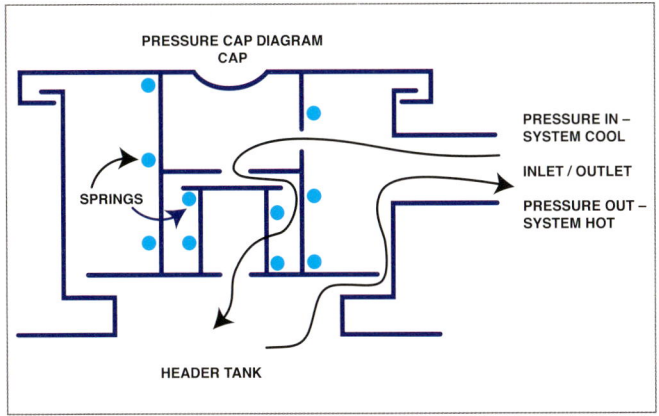

A typical pressure cap operation explained.

Radiator blanking cap. Used to replace the radiator pressure cap if a separate header tank is installed.

Radiator pressure cap. Releases excess pressure from the cooling system, and allows air or water back into the system as pressure falls.

HEADER (OR BREATHER) TANK AND PRESSURE CAP

Cooling systems operate under pressure, usually somewhere between 0.5-1.0 bar. This improves heat exchange in the radiator and, within limits, prevents the coolant from boiling. This is very useful – but it's also quite helpful if the pressurised system doesn't blow itself apart when the coolant heats up and expands. A header (also known as a breather) tank with a pressure cap accommodates the changes in coolant volume and system pressure that accompany changes in temperature.

A header tank is partly filled with coolant and must sit at the highest point of the cooling system, either incorporated into the top of the radiator, or as a separate pressurised tank connected to the bottom radiator hose (which is the only option if the radiator has to be mounted low down in the engine bay). As the coolant expands and contracts with temperature, the fluid level in the header tank changes, but the rest of the cooling system remains full and under pressure.

As a further refinement, high points in the cooling system that might trap air or steam pockets can also be vented into the header tank via air bleed pipes.

Excess pressure in the system is vented through the pressure cap, but as the system cools and the coolant contracts air is drawn back into the system through the pressure cap to prevent the header tank, radiator and hoses collapsing.

A separate header tank – either as an aftermarket kit or from a donor vehicle – can be added to the cooling system by replacing the radiator pressure cap with a simple sealing cap, and connecting the radiator overflow to the side of the new header tank. Some donor header tanks are equipped with low fluid level sensors that connect to a dashboard warning light. An aftermarket kit is also available.

Bottom outlet connects to the bottom radiator hose, side outlets connect to air bleeds, overflow outlet connects to expansion tank.

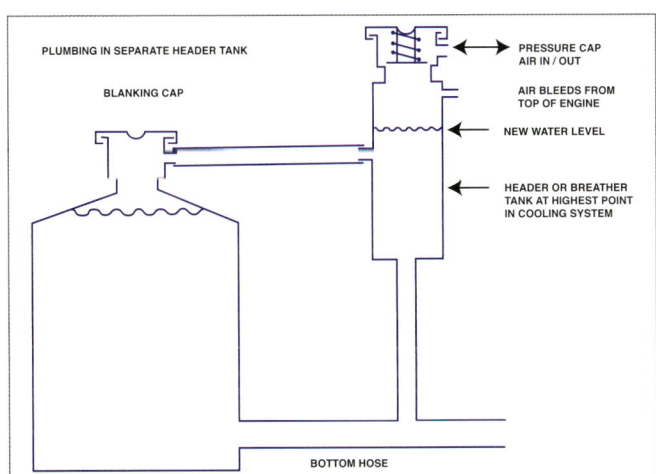

The layout of a typical remote header tank. The tank must always be located above the height of the radiator.

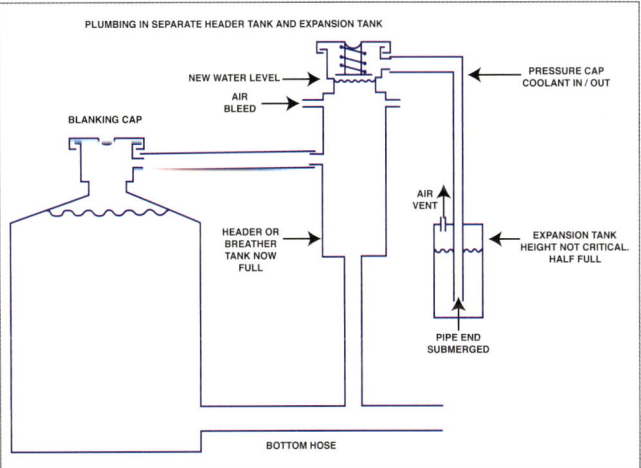

Expansion tank is not the same as a header tank. This shows how the two can be incorporated into a system.

Modify, Improve & Upgrade Your Kit Car 9

Aftermarket fans are lightweight. Make sure you have the motor spinning in the correct rotation.

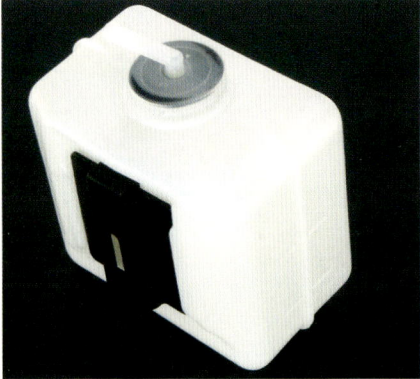

Expansion tank. The plastic tube at the top runs to the bottom of the tank, so the open end is always below the coolant level.

FAN

Thermostatically controlled fans are widely available in a large range of sizes, so can be closely matched to the radiator core size. Donor fans obviously fit the donor radiator perfectly, but can be heavy and bulky. It's important to check the direction of rotation of the blades – induction motors can be reversed by swapping the supply and earth wires, but get it wrong and the fan could draw hot air from the engine through the radiator. This is unhelpful.

The fan relay controller is usually found in the radiator or top hose – either a simple thermoswitch or a temperature sender for the ECU. Variable temperature fan switches are available, or you could bypass the sender altogether with an override switch in the cabin. Aftermarket senders will probably not match the available ports on your engine, but adapters or hose inserts with a sender welded in are available.

EXPANSION TANK

Although the terms header, breather and expansion tank are often used interchangeably, they are not the same thing. An expansion tank is not pressurised, and can be mounted at any level relative to the cooling system. It is part filled with coolant, and its inlet/outlet pipe runs almost to the bottom of the tank, so that the open end is permanently submerged. The expansion tank inlet/outlet pipe is connected to the header tank pressure cap outlet, so that coolant, rather than air, is drawn in and out of the header tank to accommodate pressure variations in the cooling system. With this arrangement the header tank is completely filled with coolant, and coolant level changes occur within the expansion tank instead. An expansion tank allows a larger variation in coolant volume to be accommodated, so is useful for larger capacity engines.

COOLANT PIPES

The layout of the major cooling system components will probably be different to the donor vehicle's, so some ingenuity will be required. Most pipe runs can be formed with either a mixture of standard pipe bends (45, 90 or 135deg made from either rubber, silicone or aluminium) and metal connectors or a continuous length of flexible silicone hose. A specialised coolant pipe cutting tool can save a lot of waste and frustration when trimming flexible pipe to size. Alternatively, flexible pipe can be fitted around a tube and trimmed to size with a scalpel. A jubilee clip can be used as a cutting guide. Pipes up to 25mm ID following a particularly tortuous route can be held in shape with spiral steel former. Slight mismatches between coolant hoses and components can be taken up using a band of bicycle tyre inner tube, or purpose made mismatch adapters. It's worth stressing that metal coolant pipes should always have a beading swaged into either end.

Silicon hose is the ideal pipework.

Stainless pipework is available with different degree corners. It must be swaged at both ends to stop hoses popping off when under pressure.

If you have a lot of pipework to fit, cutting it neatly can be difficult. These are designed for the job.

Sometimes you may need a hose reducer to join pipework to components with different diameter outlets.

Car heaters are huge, but aftermarket ones such as this from CBS are perfect for a kit.

In the engine bay

You can get a coolant sender thread adaptor if the standard thermoswitch isn't compatible.

Aftermatker radiators won't always allow for a thermoswitch. This inline housing overcomes that.

A typical thermoswitch for installation directly into the radiator or an inline application.

HEATER CIRCUIT

Donor car heaters are big, bulky and have complicated control systems. Various dinky and straightforward aftermarket heaters are available which are much more suited to compact kit car interiors. Most will also require a separate water valve, control switch and ducting. Heaters are plumbed into the coolant bypass circuit – which operates before the thermostat opens and diverts hot coolant into the radiator – otherwise the heater won't work properly until the engine reaches normal operating temperature. The heater circuit will probably need a bleed facility to remove airlocks when the cooling system is filled.

THERMOSTAT

The thermostat regulates engine temperature by controlling coolant flow to the radiator. In a cold engine, coolant circulates within the engine (in a 'bypass' circuit), which allows the engine to reach operating temperature rapidly. Once this temperature is reached, the thermostat opens, and coolant can flow through the radiator.

In its simplest form, the thermostat is a simple mechanical device that relies on some expanding wax to push it open as the coolant warms up. It sits in a housing at the top of the engine and, best of all, costs about a fiver. Unfortunately, it can be a bit slow to respond to sudden temperature changes, and when it fails it tends to stay closed, so the engine overheats.

More modern engines have the thermostat and housing in a single sealed unit, which sometimes have either mechanical or ECU driven systems to detect and respond to sudden changes in coolant temperature or pressure. When they fail, they tend to stick in the open position, so the engine takes a while to warm up, but isn't killed. These thermostats are much more expensive, and sometimes a bit bulky and poorly located for rear-wheel-drive installations.

Kits to replace and relocate the thermostat housing are available for the Ford Zetec engine from companies such as Kit Spares and Dunnell Engines.

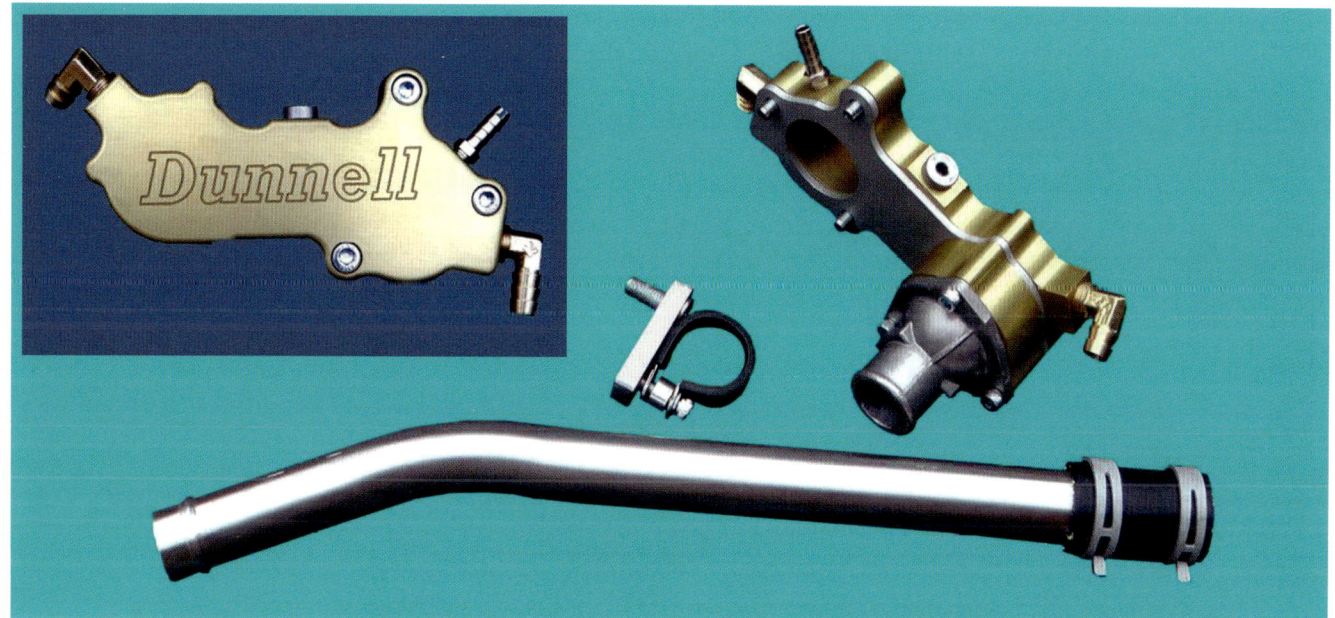

This housing incorporates the thermostat and, uniquely, replicates the Ford unit's dual function thermostat. This allows the engine to warm quickly from cold but also eliminates thermal shock of cold water entering the engine from the radiator.

Modify, Improve & Upgrade Your Kit Car 11

WATER PUMP

In most cases, the water pump that comes for free with the donor engine will be perfectly suitable, but check for a coolant leak under the drive pulley – a sign that the impeller shaft seal has started to leak and the pump needs to be replaced. Some donor engines will have their auxiliary drive belt arrangement altered, usually to dispense with power steering and air conditioning pumps. If this is the case, make sure that the correct direction of drive to the water pump pulley is maintained.

Electric water pumps have the advantage of maintaining coolant flow regardless of engine speed, which is an advantage when an engine returns to idle after a thorough thrashing. It's also possible to set them to run for a timed period after the ignition is turned off. Electric pumps are generally used in place of the original pump – the casting is retained but the impeller is removed and the pulley is not driven.

An electric water pump offers a number of additional benefits over a conventional engine driven unit.

ANTIFREEZE AND WATERLESS COOLANT

The clue is in the name, but antifreeze also contains corrosion inhibitors that are especially important in aluminium engines to prevent coolant passages filling with oxidised gunge. Antifreeze gradually loses its efficacy so should ideally be replaced every two years for traditional glycol types (blue or green), and four years for the newer organic acid (OAT) formulations (orange or pink). The two types should not be mixed.

Waterless coolant has a boiling point of around 180deg C, and does not pressurise the cooling system as is the case with water based coolants, so the system is far less likely to spring an impromptu leak. It transfers heat more efficiently, inhibits corrosion and will last for 20 years. Importantly, especially if you're a cat, it's non-toxic. However, it's considerably more expensive than traditional glycol based antifreeze.

CAR AND ENGINE SPECIFIC ADVICE

Some cars and engines have a reputation for thinking that they ought to have been a kettle, but if you're the proud or intensely frustrated owner of one, you're probably not alone. Internet forums and model specialists can be a valuable source of information, as of course, is your kit manufacturer. Beyond the standard advice, this often involves modifications such as retrofitting a header tank and suggesting suitable donors of larger radiators and fans.

With my own Minx project, which uses the infamous Rover K-series engine, the problems are a little more involved, but internet forums and specialists for the MGF, Lotus Elise and GTM describe the wholesale redesign of the cooling system that was required. Hopefully your car won't need quite as much head scratching as mine did!

Complete Kit Car would like to thank Matt Foreman of Car Builder Solutions for input into this article and for use of many of the pictures used.

Waterless coolants are expensive but work well.

An inline control valve can be connected to a control knob in the cockpit.

An inline bleed valve to help remove air in the system.

In the engine bay

Fuel Systems

John Dickens shows you the components needed to design your own fuel system.

The purpose of an automobile fuel system is to provide a constant supply of clean fuel at the correct pressure to the fuel metering components. The detailed design of the system will differ depending on whether the final fuel metering is done by a carburettor or by fuel injection, but many of the components are common to both systems.

FUEL TANK
In its most basic form, the tank is simply a reservoir containing sufficient fuel to allow the vehicle to travel a reasonable distance between refuelling stops. The most common material for production car fuel tanks used to be mild steel but moulded plastic seems to be most widely used now, allowing the designer to produce convoluted shapes which can optimise internal volume whilst fitting into previously unusable spaces.

If you intend to use a plastic tank, IVA regulations require evidence that the tank is designed for road use or is from a production vehicle and is unmodified.

Specialist tanks have often been made in aluminium alloys or GRP, but with the introduction of ethanol into petrol this latter material is falling out of favour. The number of kits still utilising the original tank from their donor vehicle is very small and most kit manufacturers offer their own alloy fuel tank with the kit or require you to have your own tank custom built.

As you can see from the illustrations in this article, the custom tank fabricators can build tanks to fit in pretty much any available space but if you do intend to have your own tank made, make sure you read the current IVA regulations covering fuel tank construction, positioning and mounting before you finalise your design Basically they require the tank to be well constructed, securely mounted and positioned away from protruding parts,
sharp edges and sources of abrasion.

They also stipulate that the tank should be earthed to prevent static build up.

There are a number of factors you need to consider when designing a fuel tank.
1. TANK CAPACITY – This depends entirely on the intended use of your vehicle. A competition or track car needs a small tank which holds just enough fuel to complete the intended race or track session. A road car should have a range of at least 150 miles and a tourer should be capable of 250 miles or more between fill-ups.
2. FUEL FILLER – You must have a large bore fuel inlet which allows fuel to flow from the petrol station pump into the tank. IVA regulations stipulate that the filler cannot be in the luggage, passenger or engine compartment which basically means that it must be on the external surface of the car or behind a flap. It must also be positioned so that any spilt fuel does not fall onto the exhaust.

The fuel filler cap must locate positively with no leaks and it must be tethered to

A competition tank designed to fit a specific space.

The tank, swirl pot, pump and filter are all contained in the boot area.

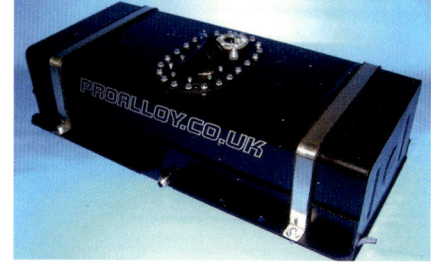
The large bore filler tube allows a remote filler cap to be fitted.

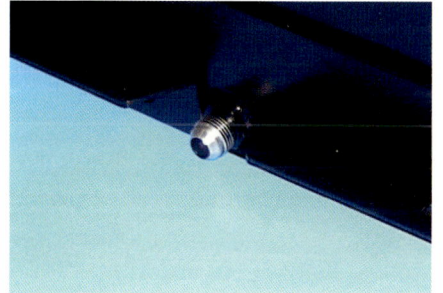
This fuel outlet uses a JIC threaded union.

An inline non-return breather valve.

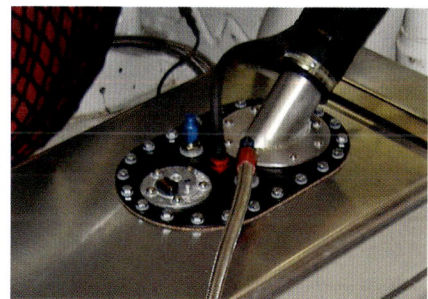
Sender unit and all unions in a removable panel.

Modify, Improve & Upgrade Your Kit Car **13**

The return pipe (and breather) are mounted in the top face of this tank.

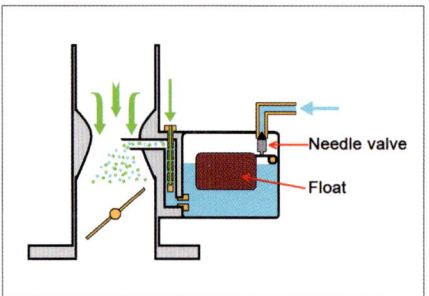
The carburettor float chamber smoothes out fuel flow irregularities.

Safety foam reduces fuel capacity only slightly but reduces fuel surge.

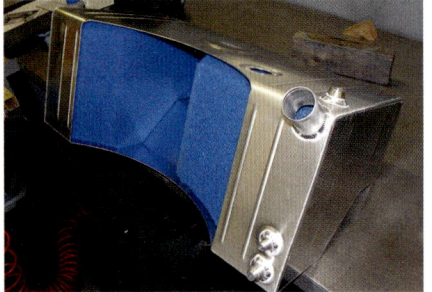
It is best fitted during assembly of the tank but some forms can be retrofitted.

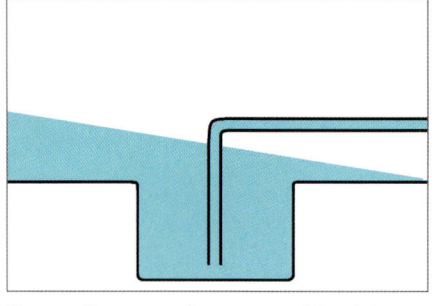
If space allows, a small sump around the pick-up can prevent fuel surge.

This central collector and baffles ensure a constant fuel supply.

the vehicle or must retain the key when unlocked and removed from the vehicle. Keep the filler pipe plumbing as simple and direct as possible. If the filler pipe offers any significant flow restriction there is the possibility that the fuel will airlock and blow back during filling.

3. FUEL TANK OUTLET – This is the small bore pipe which supplies the fuel to the petrol pump. In order to make use of all the available fuel it must pick up the liquid from the lowest possible level in the tank but, to avoid collecting any sediment which forms, it must not quite touch the tank floor. This means there will always be a small amount of unused fuel left in the tank. The outlet pipe may be a simple push fit with a hose clip or it may have a threaded JIC connection. This pipe will never have to contain any pressure as it is on the vacuum side of the fuel pump.

4. BREATHER – As fuel is drawn out of the tank, air must be drawn in to replace it or a partial vacuum will develop in the tank and the fuel will stop flowing. On early cars and motorcycles this air inlet was simply a couple of offset holes in the inner and outer fuel cap skins, but emission regulations no longer allow fuel vapour to escape into the air.

So modern tanks have a separate breather tube which vents into the air filter or intake system via a carbon filter or have a valve fitted which allows air to enter the tank but prevents petrol from escaping. This second option is also available to amateur builders in the form of a tank breather valve. It can be fitted directly into the upper surface of the petrol tank itself or it can be mounted inline in a flexible tube leading from the tank breather outlet.

IVA requirements are that the breather (or vent) should not exit into the vehicle, onto the exhaust or allow fuel to leak when the vehicle is driven.

5. FUEL LEVEL SENDER UNIT – If you are designing a tank for a road car you will also want to include a sender unit so that you can fit a fuel gauge to the car. In the distant past, out of necessity, I have driven cars with broken fuel gauges and it is not a comfortable experience. No matter how accurately you calibrate your homemade dipstick or try to work out your fuel consumption you will eventually end up running out of fuel or sucking sediment into the fuel lines.

There are two types of sender unit, the conventional float arm and the more recently introduced dip tube version which is more compact and robust but less adaptable. Whichever you choose, make sure that the resistance range of the sender is correct for the fuel gauge you wish to use. The best way to ensure this is to buy them as a pair.

If you have an engine fuelled by carburettors then a tank designed with the features described above will be fine. If you are using fuel injection on your vehicle there are a couple more factors to consider.

6. FUEL RETURN – Most modern EFI systems have a fuel pressure regulating valve which opens at a preset pressure and returns excess fuel back to the tank. The fuel tank therefore needs an additional fuel inlet pipe to accommodate the returning fuel. Since this fuel is no longer under pressure the connection can be by a push on pipe and hose clip or by a threaded JIC union.

7. FUEL COLLECTOR – Under hard acceleration, braking and cornering, the fuel in the tank will move around violently. This is known as fuel surge. If the fuel level in the tank is low, this fuel surge can uncover the fuel pick-up pipe for a short time and the fuel pump will then suck in air rather than fuel.

If the vehicle has a low pressure fuel pump feeding carburettors this is not a problem. The air pockets will simply pass through the fuel pump and into the carburettor float chamber. Since the float chamber is effectively a separate fuel reservoir anyway, it can easily accommodate a small interruption in the fuel supply and will refill to the correct level as soon as the fuel flow is restored.

An EFI system, on the other hand, cannot tolerate air in the fuel system. The high

In the engine bay

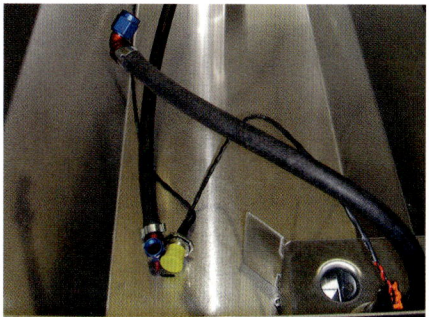
This long narrow tank needs twin collectors.

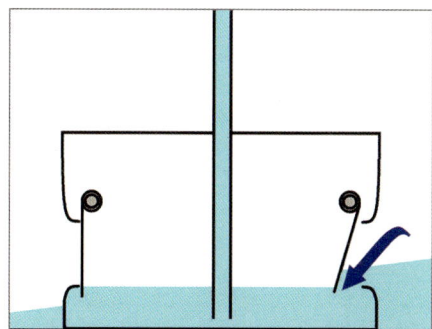
The check valves can be simple hinged flaps.

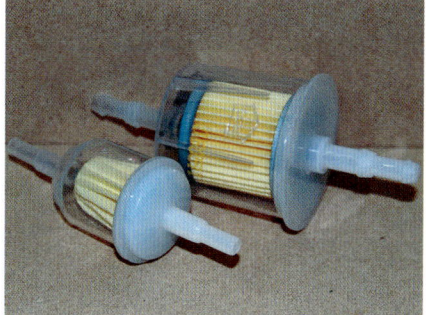
Low pressure filters are available in various sizes.

High pressure filters are usually metal bodied.

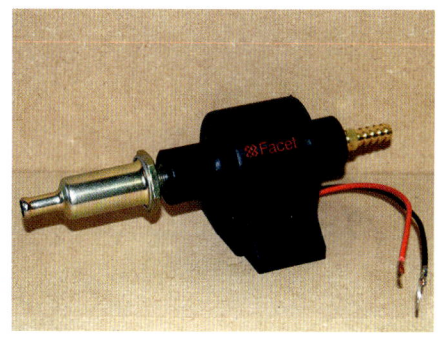
Facet Posi Flow low pressure pump with filter.

Electric pumps rubber mounted to reduce noise.

pressure injection pump can overheat and seize if allowed to run dry and any air pockets in the high pressure lines will be pumped round to the injectors where they will cause misfiring and poor running.

The use of baffles or safety foam in the tank can help to restrict fuel movement but the best way to avoid the effects of fuel surge is to build a proper fuel collector system into the tank. This may take the form of a small sump mounted below the tank which encloses the fuel pick up pipe or it may look like a small box surrounding the fuel pick up pipe. This box is fitted with simple check valves which allow fuel to flow into the collector but prevent it flowing out again. Either of these systems should ensure that the fuel pick up pipe is always submerged in fuel regardless of the forces acting on the car.

FUEL FILTERS

A brand new fuel system should start off perfectly clean, but unfortunately it does not stay that way. The internal surfaces of the tank can slowly corrode. Dirt may enter the tank when fuel is added and the fuel itself can contain solid particles picked up from storage tanks and transport vehicles.

The only way to prevent this solid matter from entering the fuel system is to filter it out. There is an argument to be made for the use of a coarse gauze filter or strainer on the pickup pipe inside the tank to remove the larger particles before they enter the system but this is only practical if the strainer can be accessed for cleaning should it become clogged.

A better option is to mount a replaceable filter between the tank outlet and the fuel pump as close to the tank as possible. This should be a reasonably fine filter but should not restrict the flow of fuel into the pump. If a finer (more restrictive) filter is needed this should be fitted between the fuel pump and the carburettor or injectors. If the filter is to be fitted into a high pressure fuel line it must obviously be a high pressure filter designed for fuel injection systems.

FUEL PUMP

Low pressure pumps – For over 80 years motorcycles relied entirely on gravity for their fuel feed since the petrol tank was always mounted above the carburettor. Although early cars used the same system, the necessity to provide adequate luggage and passenger space meant that petrol tanks were moved below the boot or floor, requiring a pump to be fitted to raise the fuel back up to the carburettor float chambers.

Carburettors need their fuel supplied at quite low pressures. Usually 2 to 4psi is fine although some of the bigger American four-barrel carburettors need closer to 6psi. The low pressure pumps which supply this fuel can be electrical or they can be mechanical pumps driven by the engine. They normally need to be mounted at the same vertical level as the fuel in the tank as they are not very efficient at lifting the fuel or self priming when empty.

Typical electric pumps will have a maximum lift of around 300mm (1ft). Once primed, however, they are very efficient at pushing fuel around the rest of the system. Electric fuel pumps can be noisy so they are often rubber mounted for sound insulation. The other requirement for a fuel pump is that it should be capable of supplying enough fuel to meet the demands of the engine when it is running constantly at full throttle. This requirement is dependent on a number of variables, such as the power output and the brake specific fuel consumption of the engine. The accompanying table gives a rough guide to the fuel flow required by a typical piston engine at different power levels.

Bhp	l/h	g/h
50	15	4
100	31	8
150	46	12
200	61	16
250	77	20

HIGH PRESSURE PUMPS – Electronic fuel injection systems with solenoid triggered fuel injectors typically operate with fuel pressures of around 3bar (45psi). This is normally generated by an electrically powered high pressure pump.

On production cars this pump is often mounted in the petrol tank itself for cooling purposes. The drawback to this design is that a large amount of external pipework then contains high pressure fuel. If the high pressure pump is mounted close to the injectors only a small section of the external fuel system is under high pressure.

Wherever the pump is mounted it must be capable of moving enough fuel to keep the engine supplied when running constantly at maximum power. High pressure pumps tend to run hotter than their low pressure counterparts and should be kept away from heat sources if at all possible. They use the fuel flowing through them for cooling and should never be allowed to run empty as the heat build up can cause permanent damage.

Finally, for safety, electric pumps must be wired in such a way that they switch off if the engine stops running. This will prevent fuel from continuing to flow if a fuel line becomes disconnected or fractured in an accident. Modern cars integrate this feature into the ECU but it can also be achieved by powering the pump through a relay linked to the alternator warning light or the oil pressure switch.

FUEL PRESSURE REGULATORS

CARBURETTOR – The fuel level in a carburettor float chamber is critical to its proper operation. It is controlled by a float and needle valve rather like those in a bathroom cistern. Excess fuel pressure can force fuel past the needle valve raising the fuel level and causing rich mixtures to be fed to the engine. If the fuel pressure remains excessive the float chamber can flood and raw fuel will leak from the float chamber vent.

To avoid this situation, a fuel pressure regulator can be fitted which, as its name suggests, will prevent the fuel pressure at the carburettor rising above a preset value no matter what pressure the pump is generating. These regulators normally work by using a spring loaded diaphragm to close a needle valve and shut off the fuel when the fuel pressure gets too high. The preload on the spring can be adjusted to set the operating pressure.

INJECTION – As we have already seen, the high pressure fuel pumps used in fuel injection systems use the fuel flowing through them as a coolant, so this fuel flow must not be interrupted or restricted. High pressure regulators therefore work as bypass or bleed valves rather than shut-off valves. They may be fitted into the fuel line before the injector fuel rail so that when they open at their preset pressure the excess fuel bypasses the injectors and returns to the tank.

Alternatively they can be located in the line after the injectors in which case they maintain pressure by allowing excess fuel to bleed back to the tank.

FUEL RAIL

In order to minimise the amount of high pressure plumbing around the engine, it is not normal practice to route a separate fuel line to each injector. Instead the injectors are linked by a single rigid tube which carries the high pressure fuel feed to them all. This tube is called the fuel rail.

SWIRL POT

If you are fitting fuel injection to a vehicle using a fuel tank which was originally designed for a carburetted vehicle or if, for any reason, it is not possible to arrange an efficient fuel collection system as described in the section on fuel tanks, fuel surge may be a problem.

There is an alternative solution in the form of a swirl pot. This is a small fuel tank of about two litres capacity which is fed by the low pressure fuel pump. From the top of the swirl pot a return line takes excess fuel

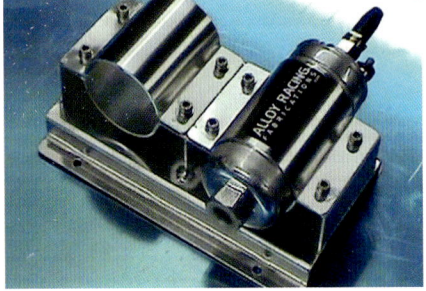

This is a high pressure pump in an Alloy Racing Fabrications mount.

An adjustable fuel pressure regulator is ideal to prevent flooding.

This popular regulator also includes a replaceable filter.

This high pressure regulator is fitted into the fuel rail itself.

This is the fuel rail on a set of Honda CBR600 throttle bodies.

A specially fabricated fuel rail for aftermarket throttle bodies.

In the engine bay

A bulkhead mounted swirl pot with JIC unions from OBP.

This swirl pot, also from OBP, is floor mounted.

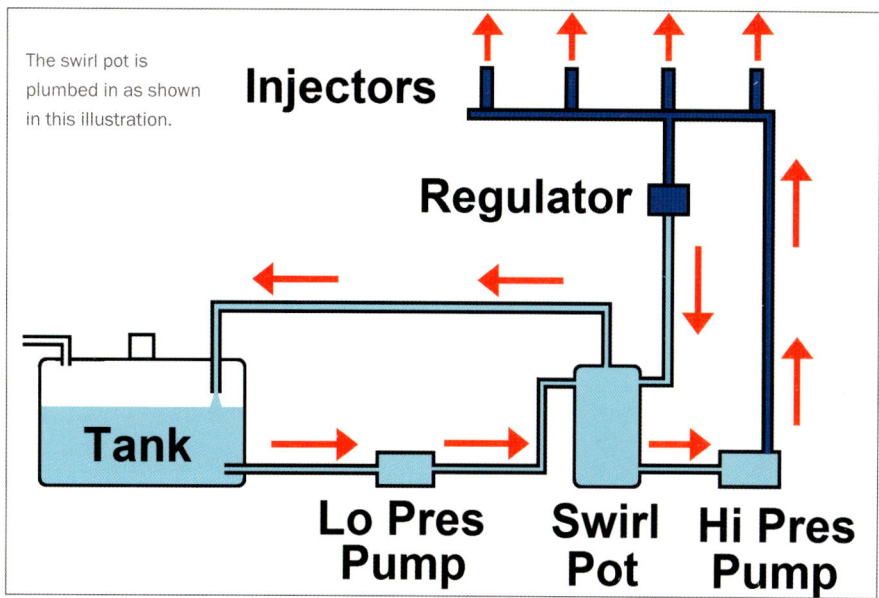

The swirl pot is plumbed in as shown in this illustration.

back to the tank ensuring that the swirl pot is always completely full. The high pressure pump picks up air free fuel from the swirl pot and excess fuel is returned to the swirl pot from the pressure regulator. This return line is angled so that any air is removed from the fuel as it re-enters the swirl pot.

In effect a swirl pot is a small fuel tank which is kept constantly full as long as the engine is running.

PIPE, UNIONS AND CONNECTORS

Fuel system components are usually linked by a combination of flexible and rigid tubing which together make up the fuel lines. The internal bore of the pipe work is normally 8mm or 10mm (5/16in or 3/8in). These lines need to be connected to each other and to the rest of the system at various points.

The injectors themselves often have their own unique connectors to link them to the fuel rail but the rest of the system normally uses either flexible tubing which is pushed on to barbed or swaged connectors and secured with proper fuel injection clips or male and female threaded fuel unions which are similar to brake unions but bigger. Commonly, these threaded unions use the Imperial JIC system or, less often, NPT tapered threads (also Imperial) but there are metric fuel unions too.

Push-on connectors are cheap and simple. Threaded unions are more secure and are more suitable for unions which need to be disconnected frequently although push-on 'quick disconnects', originally designed for motorcycles, can be fitted into low pressure flexible tubing too. These small components allow unrestricted fuel flow when connected, but seal off the lines when disconnected allowing easy servicing. I use these on my vehicles and they are very useful, but take care when you buy them as some of the cheaper versions only seal at one side. It is also a good idea to buy spare O-rings as they can swell in use, making the two parts difficult to reconnect.

POSITIONING FUEL COMPONENTS

Wherever possible, the fuel system components need to be located away from sources of vibration and, of course, heat. High pressure pumps generate heat anyway and if the fuel lines are exposed to engine or exhaust heat too, there is the possibility of fuel vaporisation, vapour lock and pump damage. Once installed, the fuel lines need to be secured with proper clips in exactly the same way as brake lines to prevent abrasion and chafing.

Designing the fuel supply of a vehicle, like designing the electrical wiring, is not difficult but needs to be done methodically and carefully with safety in mind at all times. Fortunately, a range of high quality components is readily available which makes the job far less daunting than it could be.

Thanks must go to Alloy Racing Fabrications, OBP, Pro Alloy and Europa Spares for providing the images and samples used in this article.

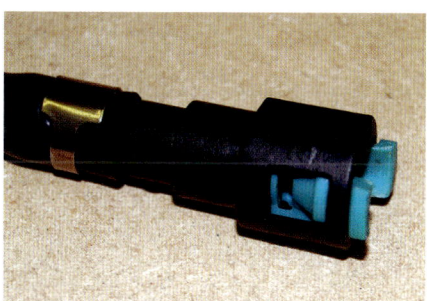

This is a typical OEM fuel connector (in this case, from a Honda).

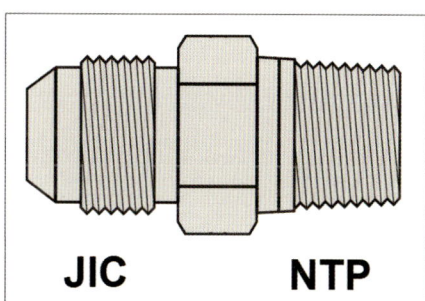

JIC unions are generally more common than NTP on vehicles.

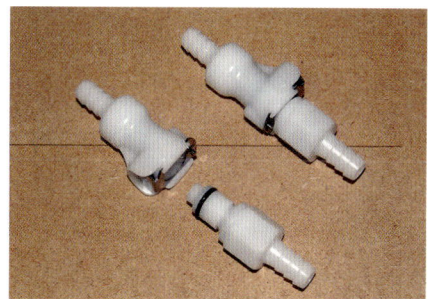

These 'quick disconnects' can be handy for servicing.

Modify, Improve & Upgrade Your Kit Car 17

Hose Clips

John Dickens outlines your options when it comes to hose clips – what you need consider for relaibility and durability.

John used a lot of hose clips on his GTM Coupé...

For high pressure applications such as hydraulic or oil lines, flexible hoses are normally secured to their end fitting using permanently crimped connections. For lower pressure hoses the normal method of securing them is to use hose clips. Not surprisingly there are a number of different types of clips depending on their intended application.

Automotive coolant hoses need to contain hot liquid at pressures up to 15psi and worm drive hose clips have been used for this purpose since the original Jubilee clip was invented by L Robinson & Co in 1921. OEM clips, particularly on British cars, were often the cheaper bent wire type but these are best avoided as they tend to distort under tension and can cut into the hoses severely.

The standard Jubilee type clip has a steel or stainless steel band with a thread formed around part or all of the circumference and a housing with a grub screw which engages with the threaded band. The thread form may be cut down into the steel band, raised up above the surface, or cut completely through the band. Personally I have found no inherent problems or specific advantages for any of these designs.

Normally hose clips are around 12mm wide but there are narrower types available at around 9mm wide. These exert a greater clamping pressure since they grip over a narrower area but they tend to do more damage to the hose if over tightened. The narrower types are often preferred for securing silicone hoses onto un-swaged inlet manifold stubs.

Choosing the correct size of clip is important if it is to clamp down correctly. The range of suitable hose diameters is normally marked on the clip itself. Unfortunately, as the hose diameter decreases, the worm drive clips become less suitable as the worm drive housing distorts the band as it tightens and it no longer forms a perfect circle. Even the specially produced mini-clips suffer from the same problem.

A better alternative, for smaller diameter hoses, is to use the screw type clips often sold as fuel injection hose clamps. Fuel

...and even more when he fitted an oil intercooler.

OEM wire clips can damage hoses – best discarded.

A standard width plated steel worm drive clip.

This thread is cut down into the band.

This pressed thread is raised up above the band.

This thread is cut right through the band.

In the engine bay

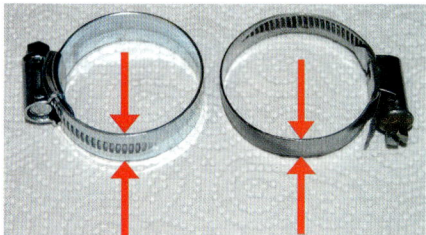
The 12mm standard band (left) and the 9mm narrow band (right).

The narrow clips can be useful where more pressure is needed.

The maximum and minimum diameters are marked on the clip.

Worm drive clips in very small diameters can distort.

This mini-clip is designed for smaller hoses but...

...even so it can distort when tightened.

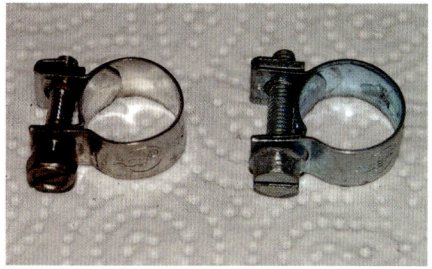
These screw clamps are better for small hoses.

The Oetiker clamp is crimped for permanent fitting.

Tiny 7mm spring clips good for carb breather hoses.

Spring clips suitable for low pressure hoses only.

Nylon clips are bulky but quick and easy to use.

This kit is to make your own hose clamp in any size.

Injection systems operate at 45 to 50psi so these connections need to be secure. These clamps are also available in plated or stainless steel. For permanent connections such as those found on fuel pumps or filters contained inside the petrol tank itself, the Oetiker clamp is a good choice. This is a crimp type clamp which is secured using the special Oetiker pliers, a pair of pincers or, if used carefully, a pair of side cutters. For lower pressure applications such as non-injection fuel lines, which operate at 4 to 6psi, spring loaded clips are suitable. Once again there is a number of types and a range of sizes. I also use the nylon 'Herbie' clips in low pressure applications. They are quickly fitted and released using pliers but the clamping mechanism is a little bulky on the smaller sizes.

There are kits available now which enable you to make up your own worm drive hose clamps. They contain a length of threaded band which you can cut to size and a number of grub screw housings which clip to the band to form the clamping mechanism. These may be a useful addition to your toolkit or your workshop for those odd times when you need a hose clamp and you just don't have the correct size in your spares collection.

There is a tendency, when fitting hose clamps, to over tighten them. This can distort or permanently damage the hose and actually cause it to leak. Just because a hose clip may have a hexagon head on the worm drive this doesn't mean that you should use a spanner or a socket driver to tighten it. Ideally a hose clamp should be tightened just enough to seal the connection and no more. It may, however, need to be re-tightened after a few days when the flexible hose may have compressed slightly in use.

Modify, Improve & Upgrade Your Kit Car **19**

Throttle Bodies

Converting a car to run on motorcycle thorttle bodies has become a popular way of affordably improving induction. **John Dickens** is your guide to how it's done.

When automobile engines used carburettors to meter their fuel/air mixtures it was normal practice for standard engines to use one or perhaps two carburettors mounted on manifolds which distributed the mixture evenly between the cylinders. This system produced reasonable performance with economy and driveability.

For maximum performance, however, it was usual to use individual carburettor chokes for each cylinder. This was normally achieved by using multiple twin or triple choke Weber or Dellorto sidedraft or downdraft carburettors. These devices represented the pinnacle of performance carburettor technology but were very expensive.

All modern car engines are now fuelled by electronic fuel injection (EFI) and although in the early days of this technology single-point injection was used it is now common practice to use individual fuel injectors for each cylinder with a single throttle body controlling the airflow into a shared air box or plenum.

Once again though, for maximum performance it is preferable to use individual throttle bodies and injectors for each cylinder. Jenvey is perhaps the best known supplier of throttle bodies but, as always with precision engineered performance parts, the investment is significant. So once again owners have looked towards high performance motorcycle engines for a cheaper alternative and are beginning to fit motorcycle throttle bodies and aftermarket ECUs to their car engines.

This Ferrari V12 engine uses six twin-choke Weber carburettors.

PROS AND CONS

Even as a confirmed Luddite I have had to admit that the carburettor versus EFI argument is all but over, as even the best carburettor systems will struggle to meet the latest emission figures whilst retaining good driveability and fuel economy. A well set-up EFI system has the ability to meter fuel far more accurately and will provide greater performance with lower emissions and much better fuel economy than any carburation set up.

Similarly, for maximum performance, the benefits of individual throttle bodies are well established and there are a few companies who specialise in providing complete bolt-on conversion kits for many common engines. Unfortunately, the cost of these kits is even higher than the cost of high performance carburettor systems and can be in the region of £2000 to £2500.

Bike throttle body conversions have become popular over the last few years as a means of achieving similar results at a fraction of the cost. Obviously such a

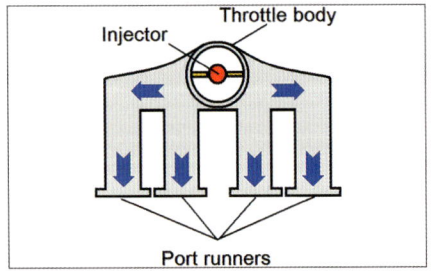
This schematic shows a single point injection system.

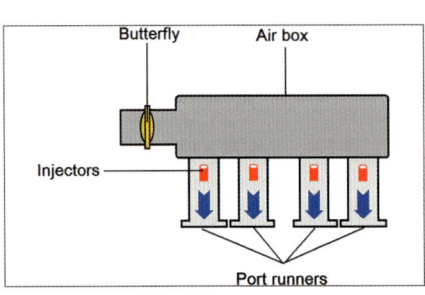
Most modern systems now use port injection with a single throttle body.

These Jenvey throttle bodies fit onto Weber manifolds.

20 *Modify, Improve & Upgrade Your Kit Car*

In the engine bay

stubs when mounted with silicone ho[ses and]
Jubilee clips. Dan has found that this
is almost always due to a combinatio[n of]
poorly aligned manifolds, loose fittin[g]
hoses, incorrectly sized hose clips an[d]
made from the wrong materials. He h[as]
found that it is possible to produce a [very]
rigid mounting if the following key feat[ures]
are observed.

- The manifold stubs must be parall[el to] the throttle bodies and have exact[ly the] same spacing.
- The hoses must be fuel resistant. [They] must be fluoro-silicone lined and c[are] must be taken not to damage this [lining] during installation. The gap betwe[en the] manifold and throttle body stubs m[ust] be as small as possible to minimis[e] the exposed length of hose and no[thing] should be allowed to contact the h[ose] outer surface.
- The hoses should be sized correctl[y] and if the outer diameters of the t[wo] components are different, the hos[e] should be sized to the smaller diameter then stretched to fit the l[arger] component. This is far more effect[ive] than trying to compress a large ho[se] onto a smaller stub.
- Narrow (9mm) hose clips should b[e used] on the throttle body stub. Typically [the] stubs are only about 10mm wide s[o a] wider hose clip will tend to squeez[e the] hose off the throttle body.

ANCILLARY COMPONE[NTS]

Fuel supply – Apart from the use of motorcycle throttle bodies, the rest of [the] components needed are exactly the s[ame] as for any other EFI system. A low pres[sure] pump is used to feed fuel to a swirl po[t and] from there a high pressure pump feed[s] filtered fuel to the injectors. Most mot[orcycle] fuel rails have a single feed pipe with [no] return outlet, so the fuel pressure regu[lator] is fitted in 'bypass' mode, bleeding off

Manifold plate and tubular runners for asse[mbly].

These early Suzuki units can be re-spaced to suit the port spacing.

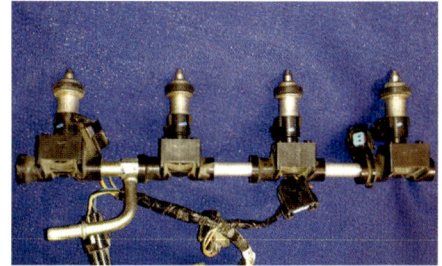
Entire fuel rail easily lifted clear.

New spacers can be used to adjust the spacing.

conversion will require more work by the individual owner/builder but a number of companies have grown up to make the process easier by providing some or all of the components needed to complete the fitment. One of these companies, DanST Engineering, kindly offered to provide images and information for this article, illustrating the basic steps involved in such a conversion.

SIZING BIKE THROTTLE BODIES AND INJECTORS

For 4-cylinder car engines the most commonly used throttle bodies are taken from motorcycle engines in the 600cc to 1400cc range. All the major Japanese motorcycle manufacturers offer sports machines in these classes so there is a wide choice.
Honda – CBR600, CBR900/1000RR Fireblade, CBR1100XX Blackbird
Kawasaki – ZX7R, ZX9R, ZX12R
Suzuki – GSXR600, GSXR750, GSXR1000, GSXR1300 Hayabusa
Yamaha – R6, R1

These later Suzuki units are paired so only partial re-spacing is possible.

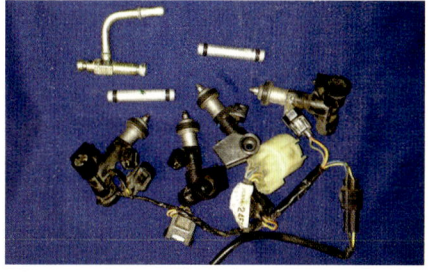
The rest of the components are push fittings.

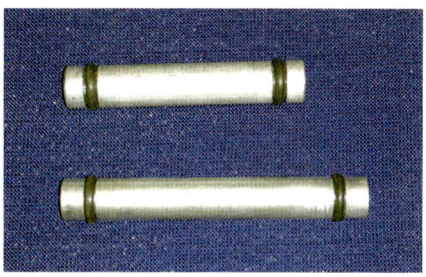
Standard (top) and extended fuel rail connectors.

In terms of internal diameter, the common throttle body sizes are 38mm, 40mm and 42mm. The most important consideration in choosing the correct throttle body size is to select them from a motorcycle engine with a similar power output to your car engine, not a similar capacity. Typically a 100bhp 2.0-litre Pinto engine could use 38mm throttle bodies whereas a 150bhp 2.0-litre Zetec engine will probably need 40mm or 42mm versions. For even larger power outputs there are 48mm throttle bodies available from the Suzuki GSXR1300 Hayabusa and the Kawasaki ZX12R.

In general though, bigger is not always better and selecting throttle bodies which are too large will result in poor throttle response due to low gas speeds and poor fuel mixing.

SELECTING THROTTLE BODIES

The early Suzuki GSXR range is popular due to its simple design and the fact that the throttle bodies are made up from four individual castings allowing them to be

Try to get all the ancillaries when you buy the throttle bodies.

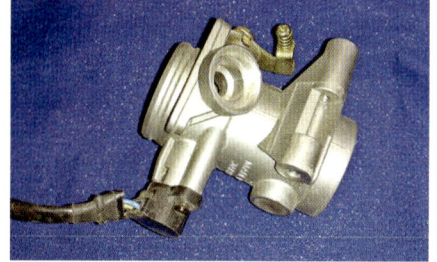
Throttle bodies are simple castings with no jets.

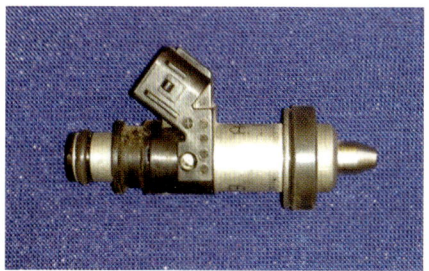
The injectors may need to be cleaned ultrasonically.

re-spaced to suit the car engine inlet ports. Later versions are cast as pairs but some re-spacing is possible allowing equal length manifold runners.

Other units commonly used are from the R1, CBR900RR and ZX12.

The injectors themselves need to be high impedance (at least 10 to 12Ω) in order to work with most aftermarket ECUs, but most motorcycle units are suitable in this respect. The injectors from the bigger bikes have flow rates of around 240cc/min which should be good for 200bhp or so without exceeding the recommended duty cycle.

It is also possible to fit injectors with higher flow rates from some Japanese cars.

CHECKING SECOND-HAND THROTTLE BODIES

In general the condition of motorcycle throttle bodies is usually good as the majority of sports bikes cover very low mileages, typically less than 2000 miles per year. When you buy the throttle bodies make sure that they are undamaged and come complete with all their to a small reservoir which collects the used brake fluid. Sealey also produces the VS020, a vacuum system which is operated from a standard compressor. It uses the venturi effect produced by the compressed air passing through the tool to generate a vacuum in the reservoir.

The bleeding operation is then controlled by a trigger. It is also available as the VS021 kit with an additional fluid container which automatically keeps the fluid reservoir topped up. The VS020 costs £25 to £35 and the VS021 is slightly more expensive at around £45, but both these units also need a compressor to operate them.

I have used both pressure and vacuum systems on my own vehicles and my personal preference is for the latter. This is not because it is any more efficient but simply because I am always a little wary of having containers of pressurised brake fluid in and around my vehicle.

PREPARATION

It is possible for air to be sucked in past the bleed nipple threads when using a vacuum pump. This air cannot enter the braking system but it can reduce the efficiency of the bleeding process so it is a good idea to seal the threads with some PTFE tape as used by plumbers. Two or three of layers of tape should be sufficient but make sure you wrap the tape in the direction shown so that it does not unwind as the nipple is screwed back into the cylinder. Make sure that all the bleed nipples and the threaded unions are tight and the fluid reservoir is topped up with new fluid.

BLEEDING

If you have a workshop manual which applies to your specific braking system it may tell you the order in which the system should be bled. If not, start with the longest line (usually the nearside rear wheel) and work your way down to the shortest line.

Remove the dust cap and connect the bleed tube to the bleed nipple but do not loosen the nipple yet. Operate the pump a few times to generate a vacuum in the system. Sealey suggests a figure of 21 in Hg. Open the bleed nipple quarter to half a turn and the vacuum will draw out some fluid and possibly air. Close the nipple and repeat the pumping and bleeding sequence until clean air-free fluid is being drawn into the tube.

Check the fluid level in the master cylinder reservoir regularly during the process and

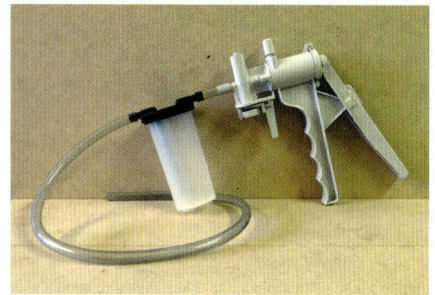
The Mityvac was one of the earliest manual vacuum pumps.

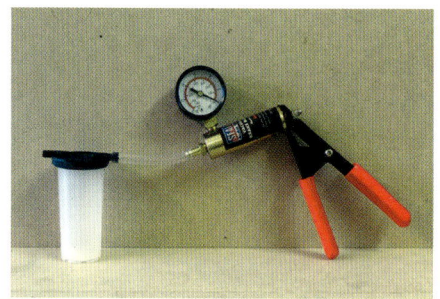
The Sealey VS402 has an integral gauge which is useful.

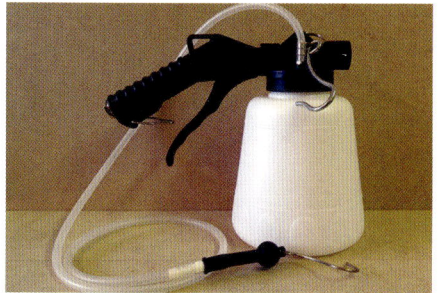
This is John's Sealey VS 020 compressor driven vacuum bleeder.

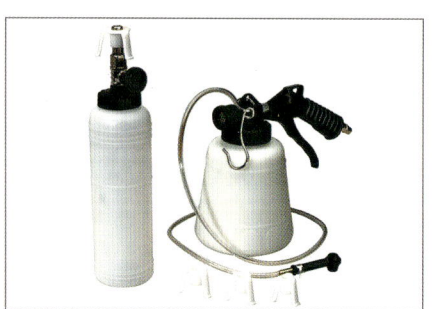
The Sealey VS 021 kit also includes a unit for topping up the fluid reservoir.

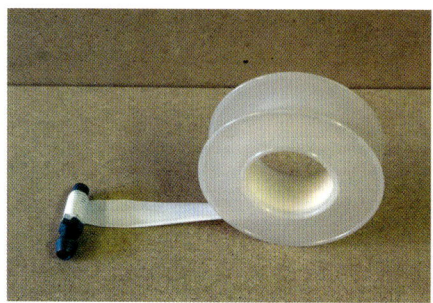
Plumber's PTFE tape can be used to seal the bleed nipple threads.

Before bleeding the brakes, check all the unions and fittings.

Ensure fluid reservoir is topped up before you begin.

Connect the bleed tube to the bleed nipple.

Operate pump to generate a vacuum in the system.

Brakes

Open the bleed nipple and watch the fluid as it is drawn out.

Initially the system is full of air which appears as large bubbles.

Eventually air free fluid will be drawn through and the line is bled.

This pump reservoir is quite small and will need to be emptied regularly.

Tiny bubbles like these may not be coming from inside the brake lines.

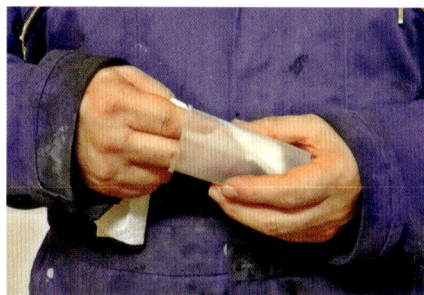
The equipment needs to be cleaned out after every use.

If possible, John prefers to use his compressor driven vacuum pump.

Simply squeezing the trigger draws out the air or dirty fluid.

top up if necessary. Also check the fluid reservoir on the pump and empty some fluid out if it is more than half full. Brake fluid should not be allowed to enter the pump itself. It may be that, after a few repeats of the sequence, only tiny bubbles are appearing in the tube. These may be the remnants of air in the system, in which case they will disappear after further bleeding. Alternatively they may be due to air being drawn in around the bleed nipple or they may be caused by cavitation as the fluid is drawn quickly through the tiny drillings in the bleed nipple. If you are happy that they are not coming from within the system itself, they can be ignored.

When you are satisfied that all the air has been removed the bleed nipple can be fully tightened and the bleed tube can be removed. Refit the bleed nipple dust caps to prevent road dirt and corrosion blocking the drillings. When all the lines have been bled, the flexible tube and the fluid reservoir on the pump can be cleaned out using an aerosol of brake cleaning solvent and paper towels.

If you have a compressor you may prefer to use one of the air powered bleed tools. This is my preferred method as, once the equipment has been set up, the process is very quick and efficient. The preparation needed is exactly the same but the bleeding process is slightly different as there is no need to manually build up the vacuum in the tool. The bleed pipe is connected to the nipple as before and the bleed nipple is opened quarter to half a turn, then the trigger on the tool is pressed to draw out the fluid and air. When the fluid is air-free the nipple can be tightened and the tube removed. The fluid level in the master cylinder reservoir still needs to be replenished regularly but the fluid reservoir on the tool is generally much bigger and is not likely to need emptying until the job is complete.

When all the brake lines have been bled, press the brake pedal a few times to make sure the brakes are operating correctly and check the system for leaks which may appear under the high braking pressures. The used fluid, even if brand new, should be disposed of at your local recycling centre. It cannot be re-used as it has been aerated during the bleeding process.

Any unused fluid can be retained for topping up once the brakes have bedded in and been adjusted, but the remainder should then be disposed of too. Brake fluid is hygroscopic and will absorb moisture from the air over time rendering it unusable. This gradual deterioration also means that the brake fluid in the system should be changed every two years using new fluid from a sealed container. With this servicing requirement in mind, I believe that these tools, retailing between £20 and £35, represent an excellent investment.

Modify, Improve & Upgrade Your Kit Car **27**

the spherical bearing may have to run at a variable angle as the pedal moves to accommodate the different master cylinder stroke lengths. However, the angulation must not be so pronounced that it causes the threaded bar or master cylinder clevises to bind on the bearing retaining tube.

The clearance between the clevises and bearing retaining tube should be around 3mm on each side. A large washer is fitted between each clevis and the bearing retaining tube, which should remain loose during normal operation, but if one circuit fails the washer will cause the pedal system to bind and retain some braking power in the remaining circuit. This is shown in the series of photos

Once the brake bias is optimised, the IVA regulations require that the balance bar is vandalised to make it non-adjustable by fully welding the locknuts to the threaded bar, and welding up any remaining thread.

Wouldn't you rather just fit a nice, simple tandem cylinder? Tandem cylinders are simpler to install, but obviously don't allow variation in front-rear brake bias. Failure of one circuit will increase brake pedal travel, but the remaining circuit will still function. Their length, and often their bulky integral reservoir, can complicate the packaging of mainstream original equipment cylinders, although Fiat made a handy 0.75in bore tandem cylinder that uses a remote reservoir. Aftermarket tandem cylinders, with either integral or separate reservoirs, are available from Wilwood, Tilton and AP.

Girling-type master cylinders use a pushrod with a mushroom end that is held in place with a circlip. Longer pushrods are available separately if necessary. Mainstream master cylinders used without a servo will need a new pushrod. This can be made very simply from a long M8 bolt with the head cut off and the stem shaped to match the original pushrod from the servo. This type of pushrod is not positively attached to the master cylinder, so the brake pedal will need to have a simple stop mechanism to hold it in the rest position. All pushrods should have a very small amount of play when the pedal is at rest.

Master cylinder reservoir: Reservoirs can be remote, and connected to the master cylinder by a pipe (which should be supplied with a declaration that it is the correct specification for brake fluid for the IVA inspector), or attached directly to the master cylinder. The reservoir must be mounted higher than the brake calipers or drum brake slave cylinders to maintain a slight fluid pressure gradient between the two. Without this, the caliper piston seals or brake drum return springs tend to draw the friction surface away from the disc or drum, so the next time the brake is applied, pedal travel is increased. If the reservoir can't

Assembly in pedal.

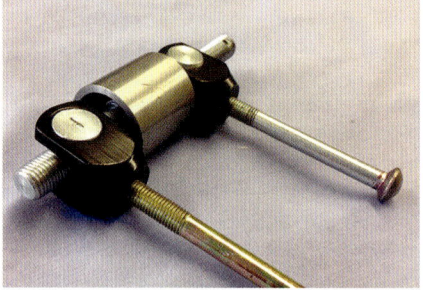
Full set shown without pedal.

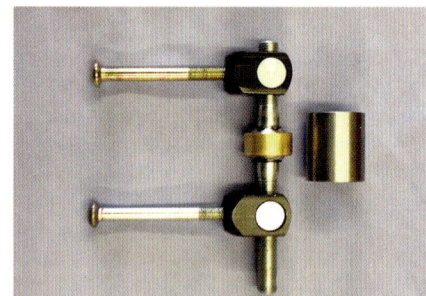
In the neutral position...

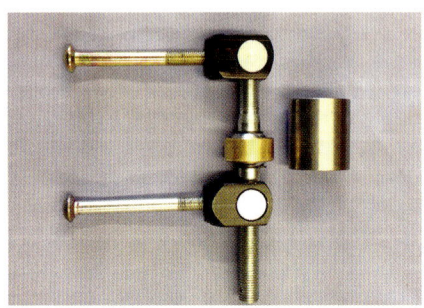
...biased to the left...

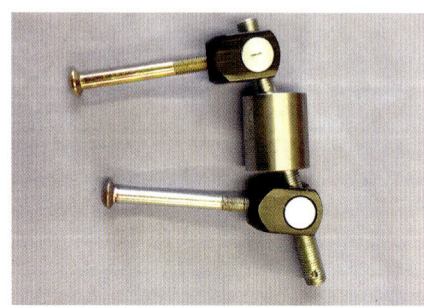
...and crooked.

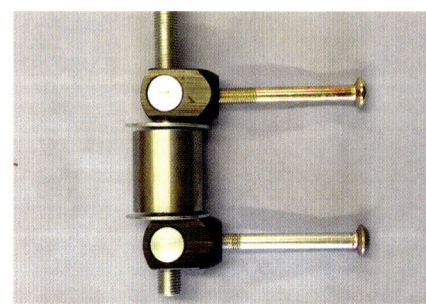
This shows it binding...

...it should have a slight gap for some play.

Pushrod...

...for the master cylinder.

30 *Modify, Improve & Upgrade Your Kit Car*

Brakes

be mounted high enough to eliminate the problem, a residual pressure valve in each brake circuit (10lb for drum brakes, 2lb for discs) can give the same effect.

A master cylinder reservoir must have a low level warning light that can be tested from the dashboard – often this is combined with the handbrake warning light, so that it's tested each time the handbrake is applied. The minimum fluid level has to be clearly marked on the side of the reservoir, and a notice indicating the type of brake fluid used in the system must be placed within 100mm of the filer cap.

Brakes pipes: Running hard brake lines can be an immensely satisfying and creative exercise. Or, it can be a complete pain if, like me, your obsessive-compulsive disorder level isn't matched by your ability, because hard lines can look truly awful if done badly. Kunifer (copper-nickel alloy) pipe is now used almost universally; it's slightly more awkward to shape than copper, but much more resistant to work hardening and cracking once in use. Copper pipe can be useful for making patterns, however.

Kunifer pipe usually arrives in a roll, but it can be straightened by pulling it through a 5mm hole in a block of wood. To avoid kinking the pipe, bends should be made with the correct tools – they're widely available, cheap, and easy to use. It's also worth investing in a small roller pipe cutter, as square burr-free cut ends form much better flares.

Hand held flaring tools work reasonably well, and are the only option for making flares with the pipe in position on the car, but using a proper, vice mounted flaring tool is an almost religious experience. If you can stretch to one, I'd heartily recommend it.

It sounds obvious, but brake unions need to be fitted over the pipe before the flares are made, and on the correct side of any bends. Most people get this wrong at least once, though...

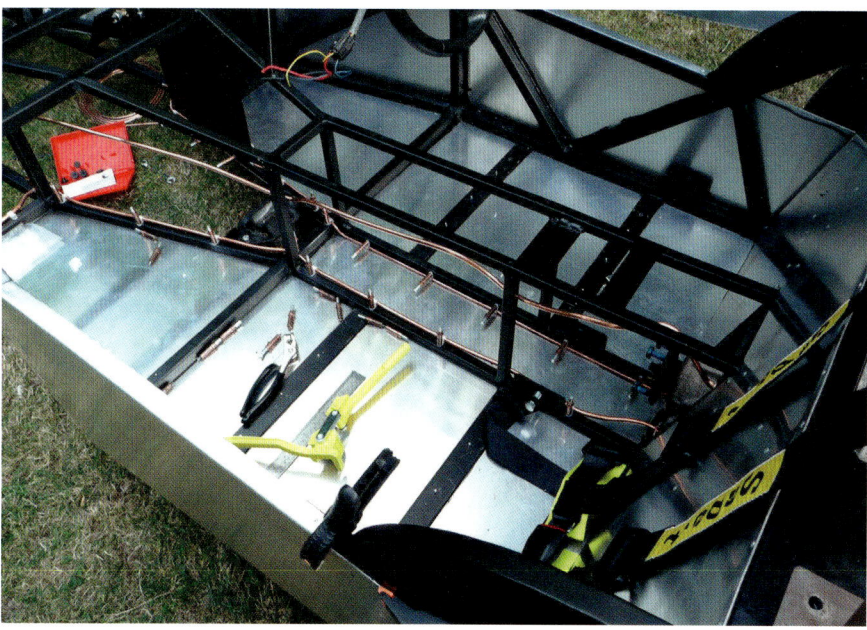

Brakes pipes need to be neatly routed and rigidly fixed at least every 200mm.

Flexible pipes are needed to join the hard lines to the brake calipers, and stainless braided pipes are a useful upgrade, as they eliminate the pressure-sapping bulging that can occur with original equipment rubber pipes. Long braided pipes are a viable alternative to kunifer pipes, if you really can't face all the bending, flaring and anxiety about leaks, or you really need to have colour coded brake lines. The master cylinder and caliper connections are stainless steel, so are much less susceptible to over-tightening than flared kunifer unions, and the solid line to flexi connections found with traditional hard line systems are eliminated. These lines work particularly well in exoskeletal-all-gubbins-on-display style cars.

Stainless lines can be made up at home using Euroquip or Goodridge fittings, if you're brave, but several companies will make them up with swaged unions from your measurements. It's worth remembering that each line needs at least one swivel, banjo or bulkhead union, or you won't be able to install it.

Brake lines should be solidly mounted every 200mm using plastic or rubber lined P, press-in or bolt-together clips. Aluminium panels don't count as a solid mount, as they can vibrate, which will promote work hardening and cracking of kunifer pipes. Obviously, brake lines shouldn't be run under a chassis rail, in case they're torn off when you're not looking. A pressure switch to operate the brake lights may be required, unless you use a pedal activated switch as in most modern production cars. This is usually fitted in to a T-piece in the brake line, but a nifty banjo bolt switch is available that saves making an extra potential leak point in the system.

Calipers: Donor calipers look pretty uninspiring when they're first extracted from the car, but they're free, they fit, they're a suitable specification, pads and discs are probably easy to get hold of and various reconditioning companies can make them beautiful again. OE manufacturer upgrades might be available, for instance the AP four-piston calipers fitted to some special edition MGF Trophys.

Aftermarket calipers can give improvements in braking performance, fade-resistance and unsprung weight. But, unless the installation comes as a complete kit for your specific car (and sometimes not even then) there can be pitfalls.

You'll need flexible lines as well as hard ones.

You can improve looks of donor calipers with effort.

Almost every kit car will use discs at the front and many will use drums at the rear...

...You need to be careful when stripping and rebuilding drums. Lots of room for error!

Most kit cars do without servo assistance, but you might want to consider it.

Larger calipers may require a larger diameter master cylinder to keep the pedal stroke reasonable, which may in turn increase the pedal effort required. To avoid this, a caliper with a total piston area similar to the calipers being replaced should be selected. Total piston area equals piston radius squared multiplied by 3.142 multiplied by the number of pistons in the caliper, although single-piston sliding calipers are treated as having two pistons.

Aftermarket calipers usually have pistons on either side of the brake disc, as apposed to many standard original-equipment 'sliding' calipers, which have one piston mounted on the inside. This means that the aftermarket caliper sticks out further beyond the wheel hub, which can cause problems with wheel clearance, particularly if you use retro steel rims.

Mounting brackets, usually made from simple aluminium blocks, will be required to adapt the calipers to the mounting lugs on your donor uprights. Radial mounted calipers, where the mounting bolts run at 90deg to the more usual lug mount bolts, give a bit more flexibility and make caliper positioning more straightforward.

Service items, such as pistons and seals, are available for most aftermarket calipers,

Pedals are either floor-mounted or pendulum. Aftermarket pedal boxes available, or you can make one!

so they can be rebuilt at home. However, pads are considerably more expensive than mainstream manufacturers' items. Some lightweight aftermarket calipers don't have dust seals on the caliper pistons; so will need inspecting and cleaning regularly. Apparently, this is because dust seals can deteriorate and cause binding in a racing environment, which is fair enough, but possibly not relevant to normal road use.

Rear calipers usually have to include a handbrake system, so once again the donor car parts make a lot of sense. However, lightweight aftermarket calipers that include a handbrake mechanism are available, as are mechanical spot calipers and mounting brackets, which can be used to adapt front-wheel-drive systems that have been moved into a mid-engine location.

Brake discs and pads: Brake discs are usually cast iron, and potentially very heavy. Larger diameter discs can be replaced with much lighter two-piece aftermarket aluminium items. These use a standard replaceable rotor bolted to a central bell that can be adapted to a specific application.

The use of ventilated discs to improve cooling, at least on the front brakes, is almost universal. Drilling and grooving of brake discs, however, remains controversial. Advocates suggest that it keeps the pads clean, reduces build up of gases between the pad and disc, and helps to dissipate water. Cynics retort that it accelerates pad wear for no useful purpose, and is strictly for silly boys in Corsas who are trying to impress bored looking girls in supermarket car parks. The grooves or line of holes are designed to run backwards from the centre of the disc to the edge, although it's surprising how often they don't!

Performance brake pads are widely available and are a worthwhile, cost-effective upgrade. They have a higher coefficient of friction and greater resistance to fade than standard OE pads, and fast road versions manage to achieve this without compromising performance from cold.

Drum brakes: Drum brakes are a perfectly acceptable choice for back brakes in a moderately powered, lightweight kit car. However, they are more complex than disc systems, with lots of scope for mixing parts up from side to side and assembling something that looks superficially correct, but won't work properly. Lots of photographs taken at disassembly will help, but it's safest to cross-reference with a manual, as the donor car may not be correctly assembled to start with.

Most systems will have a leading and trailing shoe arrangement, with a thicker leading shoe towards the front of the car. The friction material on each shoe is usually applied asymmetrically, with one end of each shoe left uncovered. On the leading shoe the uncovered end is usually next to the wheel cylinder, at the top of the brake backplate, with the trailing shoe the other way up.

Painting the brake drums satin black will improve heat dissipation, but other than assembling them correctly the only other upgrade for rear drum brakes is to swap them for discs and calipers.

Servos: Servos are not widely used in kit cars, even a snorting Ultima GTR manages quite well without one. Adding a servo to a braking system will not increase the system's maximum braking power, only reduce the pedal effort required to achieve it. That said, servos allow the use of wider bore master cylinders that the average driver would struggle to operate without using a very long pedal travel (and maybe some anabolic steroids), which are useful for large 4 and 6-piston calipers. The purist might argue that servos remove a degree of pedal feel.

Most servos rely on a vacuum supply to operate, either from the inlet manifold in a normally aspirated petrol engine, or

Brakes

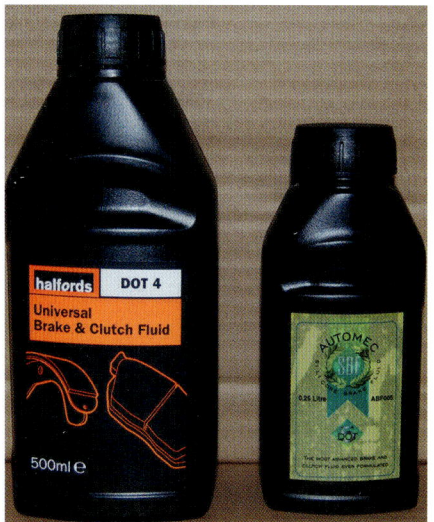
Choose the right fluid, and don't mix the wrong ones!

Brake bleeding is made a lot easier with the right tools.

via a pump with diesels or turbos. The traditional OE vacuum servo only operates up to moderate braking pressures, anything beyond that is down to the driver.

Donor vehicle servos are quite bulky, and live behind the tandem master cylinder, so they can be a bit awkward to package in a confined kit car pedal box Single-circuit remote aftermarket servos are available from classic car suppliers, that give a boost ratio of between 2 to 3:1, but are not useful for IVA compatible (or safe...) dual circuit braking systems. However, Car Builder Solutions can supply a remote dual-circuit servo kit, and ABS Power Brakes America produces a system using a new master cylinder that is pressurised by an electric pump and accumulator.

Brake bleeding: Several years ago, after messing about with pumping pedals, Gunsons' Eezibleeds and a vacuum bleeder, which was completely useless, I made a pressure bleeder out of a garden spray bottle and a spare master cylinder cap. The result was a revelation and I was enormously pleased with it. Then I discovered that Sealey sell a similar item, which took the wind out of my sails a little bit, but I'd still recommend that you buy one. They make a potentially frustrating and tiresome job a 20-minute doddle.

The bleed nipple has to be at the top of the caliper or brake cylinder for bleeding to be effective. This sounds pretty obvious, but occasionally manufacturers (in their wisdom) will mount brake components in non-standard locations, so they have to be removed to be bled correctly. Aftermarket calipers often have four bleed nipples, so they can be mounted either way up. Only the top two need bleeding.

Brake fluid: Brake fluids are either glycol based (DOT 3, 4 and 5.1) or silicone (DOT 5). DOT ratings refer to the boiling point of the fluid – the higher the better, but there are many other differences between the types, which can cause some controversy.

Glycol based fluids are reassuringly familiar, but they have problems. They are hydroscopic, in that they absorb water from the environment, which then diffuses through the system, causes corrosion and lowers the boiling point of the fluid. Opening the fluid reservoir a few times a year can introduce a one percent moisture content to the system that will reduce the fluid's boiling point by 100 deg F. Because of this, fluid manufacturers recommend replacing brake fluid every two years, not that anyone does. Glycol fluid also strips most types of paintwork – so it's worth having some water, some rags and a plan to hand before you start using it, just in case.

Silicone fluid doesn't absorb water, so never needs replacing, and it won't attack paintwork if it's spilled. So why doesn't everyone, including major manufacturers, use it?

Silicone fluid isn't compatible with ABS systems, so it's not much use for new cars, it's much more expensive than glycol, but its main problem is that it gives a 'soft' brake pedal in comparison to glycol based fluids, which gets worse as the fluid heats up. Some drivers find this very disconcerting, whereas others can live with it in a steadily driven road car. However, most aftermarket brake manufacturers don't recommend silicone fluid for track day use.

DOT 3 and 4 glycol fluids can be mixed, but aren't compatible with DOT 5 (silicone) or DOT 5.1 ('synthetic' glycol). Changing from glycol fluid to silicone is an involved process. Residual glycol will combine with the silicone to form a gel, and residual water won't be absorbed into the brake fluid, but will pool at the lowest point in the brake system causing corrosion and possibly brake failure if the water boils or freezes. The system must be thoroughly flushed through, and ideally all the rubber parts, which can harbour water and glycol, should be replaced. The silicone/glycol decision really should be made before the system is filled for the first time.

Glycol fluid should be bought in small quantities. Once the pot is opened, it will start to absorb water and go off, so storing any that's left over isn't terribly useful. Silicone fluid should be poured down the side of a funnel rather than directly into the reservoir, as it tends to form and trap bubbles very readily, which will worsen an already soft pedal. Leaving it to stand for 24 hours before use will allow any bubbles that have formed during transit jiggling to escape.

ABS: Contrary to popular belief, ABS is permitted by IVA, but if fitted it must work correctly, and have a warning lamp that lights when the system is operational. The system is usually assessed with a driving test, unless the ABS warning lamp lights during the roller speedometer test, which indicates that the system is functioning correctly.

DIY GRP Panels

Martin Scott explains how to easily make a flat GRP panel – ideal for a Seven's dashboard, a doorcard or any similar function.

Here's a method of making GRP panels without having to spend a lot of time or expense on making a plug and taking a mould off it before you can start moulding. The technique uses easily available items to create a female mould, from which the finished item is produced.

Here we will produce a dashboard for a Seven type kit, but the process can also be used to make a number of parts such as a doorcard, or fixed inner panelwork. I've used a piece of board 4ft by 15in (1120mm by 450mm) and 15mm thick.

MATERIALS AND TOOLS NEEDED

- Piece of Melamine faced shelving board 1220mm by 450mm
- Piece of planed timber 44mm on one side
- Two pieces of planed timber about 50mm by 20mm
- One piece of planed timber with one side about 4mm smaller than shelving board thickness
- All timber about 1250mm long
- Mould release wax
- Resin A and resin B
- Catalyst
- Laminating brush 50mm
- Laminating roller 9in x ¾in
- Latex or similar gloves
- Stirring sticks
- Mixing pot
- 40mm plastic pipe 1240mm
- 32mm plastic pipe 1240mm
- Surfacing Tissue
- Chopped strand mat
- Making the mould

Start with a sheet of flat shelving board (Conti-board is one brand) a little larger than the intended item, to allow for trimming and ease of moulding.

A smooth radius needs to be created for the lower edge of the dash for two reasons. It makes it easier to lay up the GRP matting/resin because it is difficult to get the matting to conform to tight corners, and the IVA inspector likes it that way as well. To create the smooth radius, I used a piece of 40mm plastic waste pipe, which in this case measures 39mm inside diameter and 44mm outside diameter – ideal as the IVA man requires a radius of 19mm minimum.

I made a couple of saw cuts and some wood to hold it in place against my conti-board to create the female mould. The timber pieces are arranged as shown in diagram one, our 44mm piece fitting as shown because it's the same size as the pipe. Plastic spacers are used between the timbers to align the edge of the pipe perfectly with the edge of the board, and a couple of screws and some double-sided tape keep it in place.

Once the female mould is created, it needs polishing a few times (as directed on the tin) with mould release wax and left to dry overnight. I also used a piece of 32mm plastic pipe, and have prepared this by waxing it. More about that later!

MAKING THE PART

This takes part in two stages – gelcoating and laying-up the matting/resin. Three layers of 450gsm chopped strand mat will result in our part being about 4mm thick.

GELCOAT

Mix up some gelcoat (called resin B) thoroughly using the directed percentage of catalyst hardener (usually 3 per cent) in a plastic container (I use old milk or food containers marked PP), and coat the surface

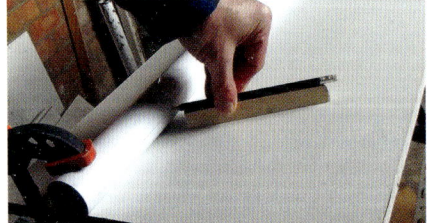
Marking the pipe that will act as part of the mould.

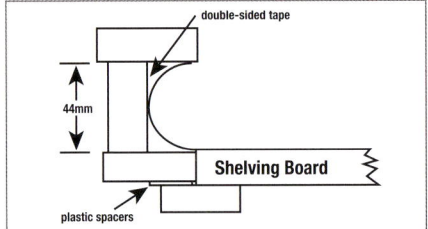
How the shelf and pipe mould is arranged.

Cutting the pipe to size.

Polished and ready to mould

Application of gelcoat first.

Leave plenty of time for the gelcoat to cure.

Bodywork & glass

Surfacing tissue applied dry...

and subsequently wetted.

Laying the GRP matting.

Second pipe adds some support.

Rolling the matting...

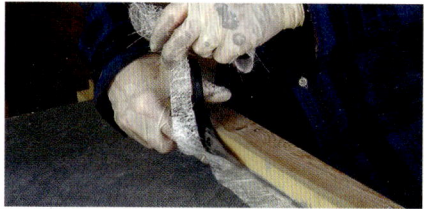
...and trimming the overhang.

generously with a brush. The intention here is to just get a fairly even layer; it doesn't need to be perfect and it shouldn't be spread out thin like paint. The amount used is about 600ml per square metre, so mix about 500ml for this job.

Gelcoat is usually coloured, and this is achieved by adding pigment or buying it ready-pigmented. Here, I've used ready-pigmented grey. After the gelcoat has 'cured', the next stage, laminate lay-up, can be tackled. If this is done too soon, it can damage the gelcoat, so a test here is to touch any thinner area (maybe where it's splashed over the edge?) and, if colour sticks to your finger, you need to wait another hour or two and then test again. As a general rule, I've found about 12 hours is OK, but this varies with temperature, and can extend to over 24 hours in cold weather.

LAMINATE LAY-UP
Here we use surfacing tissue, chopped strand mat (450g /sq m weight) and lay-up resin (resin A). Before mixing any resin, cut a piece of the surfacing tissue a little bigger than the part, and three pieces of 450g chopped strand mat also a little bigger than the part.

I've found it's best to add some colour pigment to the first lay-up resin just in case there are some tiny areas or pinpricks where the gelcoat hasn't covered well, otherwise these will be visible in the final part. If you've used ready-pigmented gelcoat, then you will need to order some pigment of the same colour when getting the materials, and mix a little (up to 10 per cent) into the resin to colour it. The surfacing tissue is 'hairier' on one side, and it's this side which faces the gelcoat.

Mix the resin (about 600ml per square metre), colour pigment and the catalyst thoroughly.

Place the surfacing tissue (it is like tissue paper) dry in position and put the resin on gently with the brush – do not use a back-and-forth brushing action (as in painting) because it will stick to the brush and create a terrible mess and you will end up swearing! Wet the tissue all over. Place the first layer of chopped strand mat in position, and use the brush and roller to eliminate air bubbles and saturate the mat (the mat will lose its white colour and almost disappear when saturated), adding some more resin with the brush to any areas that look too dry and roll again until they are saturated.

Now comes the easier bit. Mix some more resin and catalyst (but with no pigment), apply the resin with the brush and place the second layer of chopped strand mat in position and use the roller to eliminate any air bubbles and saturate the mat as before. Repeat for the next layer of chopped strand mat, and the laminating is done. If necessary, trim the mat back a little to minimise the overhang, which will tend to cause problems.

You may find the laminate doesn't want to stay in the curved part of the mould, and this is where the piece of 32mm pipe comes in. Just place this in position (with clamps if necessary) and the laminate will be held in position – the pipe I used measured 33mm outside diameter, so was a little too big to fit to the inside fully, but was near enough so that it could be pressed into position to stop the mat falling from the edge.

You may notice the laminate gets warm in the next hour, and it's advised not to laminate more than three layers of CSM in any one session. The part you have created will be about 4mm thick, so ideal for your kit car. Leave the part in the mould for at least three days to cure. The duration will be dependant on temperature.

REMOVE FROM THE MOULD
It's worth protecting your hands with some thick gloves whilst doing this, as the edge of the laminate can be quite sharp! As we are using a temporary mould, we have the advantage of not needing to keep it for further use. We can unscrew and remove the timber, and gently prise the part from the shelving board using a wooden spatula in several places. Once the part releases, it is usually an easy process (especially with a flat panel) to pull it away, and also remove our plastic pipes.

Trim the panel to size, cut holes to suit your instruments, and maybe even cover it with vinyl or leather. Making your own GRP parts and popping them out of the mould is very rewarding.

The finished panel. Just needs a trim and the mould line polishing and it's good to go!

Modify, Improve & Upgrade Your Kit Car **35**

Fibreglass Repair

Need to make a GRP repair, but don't fancy mixing resin and getting the right matting materials? Then this may be the product you're looking for. **John Dickens** explains.

Six or seven months ago, CKC Running Reports contributor Ed Morton emailed me, through the 'Ask John' facility on the CKC website, with a query about a particular GRP product. The product is essentially a repair patch using glassfibre material impregnated with a pre-catalysed UV activated polyester resin. No mixing is needed and the material is ready to use straight from the pack. The patch is simply applied to the prepared surface and cured in sunlight or by using a UV lamp.

I had heard of these products being used for emergency repairs on canoes and boats but had no practical experience of them myself. A second reader, John Lockett, sent me some follow-up information, including the product name Curon but unfortunately this seems to be out of production. Recently, however, John sent me details of another product, the Ultra V Patch from Connect Workshop Consumables, which seems to be designed to do exactly the same job.

Polyester resins are cured by a chemical reaction involving very reactive components known as free radicals. These are normally provided by the MEKP catalyst or hardener, which is stirred into the resin just before use. UV cured products use exactly the same polyester base resin but the hardener is a photoinitiator which is already mixed into the resin. When exposed to UV light this compound decomposes to release the free radicals needed to cure the resin. Obviously these UV cured products must not be exposed to UV light or sunlight before use and are always stored in opaque packaging.

The Ultra V Patch is 21cm by 11cm (8in by 4in) and is supplied in a black polythene inner envelope with full instructions printed on the outer packaging. The list price is £13.36 but I found examples discounted as low as £6.99 after a quick web search. The manufacturer claims that the patch can be used on a variety of substrates, including metals and roof tiles but obviously our main concern is its performance when used to repair damaged GRP.

The Ultra V Patch comes with full instructions.

Modify, Improve & Upgrade Your Kit Car

Bodywork & glass

Damaged GRP is best repaired from the CSM, or rear, surface (main pic on opposite page). If the rear is inaccessible the Ultra V Patch will work on the gelcoat too.

A good wash with water and detergent will remove dirt and grime.

80 or 120-grit abrasive paper will clean and key the surface.

A wipe with acetone or thinners will remove any grease from the surface.

The inner packaging is an opaque black plastic envelope.

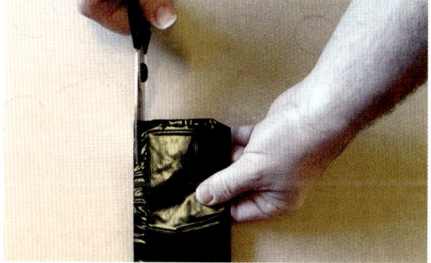
The sealed bag must be cut open to remove the patch.

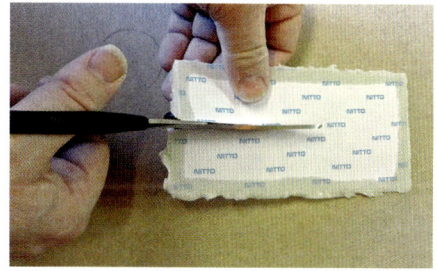
The patch itself is easily cut with either scissors or a knife.

Unused patch material must be returned to the opaque bag.

IN USE

1. The first step is to prepare the area around the damage to ensure a good bond with the patch. The instructions require the surface to be clean, dry and free from oil and grease.
   I would suggest a good wash with water and detergent to remove dirt and grime then, when the area is completely dry, it should be rubbed down with 80 or 120-grit abrasive paper to further clean and key the surface. This will also remove any loose strands of glass which could cause the patch to lift before it cures. A final wipe over with acetone or thinners (Fig 6) will remove any remaining grease and temporarily re-activate the surface resin making it slightly sticky.

2. The patch is removed from its packaging and, if required, can be trimmed to size using scissors or a knife. Any unused material should be returned to its packaging quickly or it will begin to set.

3. The patch material has white backing paper on one surface and a clear film on the other. The white backing paper is removed to expose the working surface of the patch. The material is then placed over the repair and moulded into the required shape using firm pressure to ensure a good bond. It has the consistency of stiff putty and is very sticky, so rubber gloves are a good idea here.

4. The clear film can be left in place unless the finished repair is to be shaped, sanded or painted in which case it should be removed at this stage.

5. The repair patch should now be exposed to UV light from the sun. The resin is claimed to cure in five to 40 minutes depending on the strength of the sunlight. I carried out this procedure on a sunny but cold day (8deg C) in early January and the patch cured to a useable structural condition in only 10 minutes. The repair could be bent without delaminating, separating or permanent deformation. In dull conditions, shaded areas or for a faster cure, a UV lamp can be used (Maplin sells a 75w bayonet fitting UV bulb for £4.99). Using a lamp, the resin should cure in about five minutes.

6. Once the patch is fully cured it can be sanded, filed, drilled and painted like any other GRP material.

Be aware that if the material is exposed to UV light, it will begin to set.

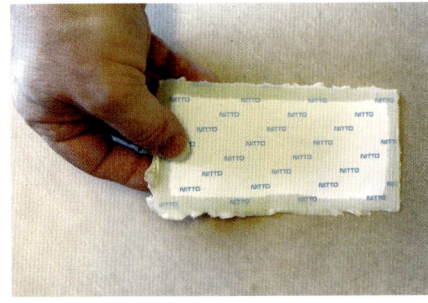
One side is covered by a white paper film. This must be removed.

The other side has a transparent plastic film which can be left on.

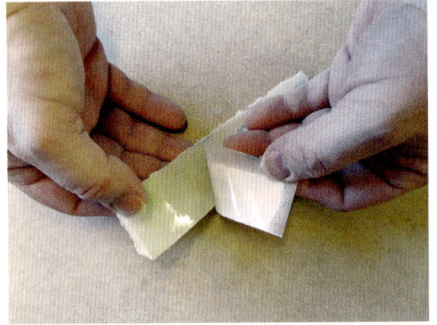
Carefully peel off the white paper film.

Position the patch over the damage then...

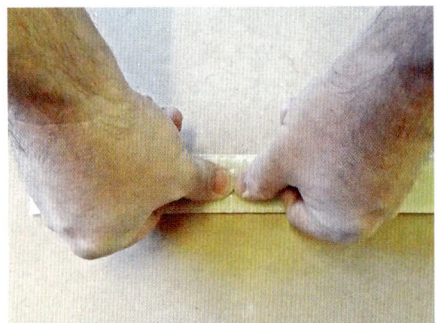
...press it firmly in to place.

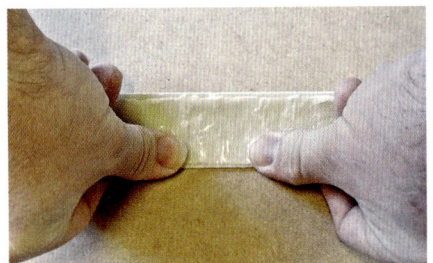
High pressure is needed to produce a strong bond.

If no further finishing is needed the clear film can be left in place.

If sanding or painting is required the film must be removed.

SUMMARY

I have to say that I was quite sceptical before carrying out this test. I had my doubts about the ability of the patch to form a strong bond to the damaged substrate and also wondered whether the patch would cure successfully in anything but strong summer sunlight. It turns out that I was wrong on both counts. The patch is firmly stuck to the GRP surface and the curing time, even in weak sunlight, is fast enough to allow additional layers to be quickly added if further strengthening is required.

Even at its discounted price, this patch initially seems expensive when compared to conventional GRP materials. A small glassfibre kit containing 250ml of resin, hardener and 1/4m² of CSM costs £9.99 at Halfords for example, but these conventional materials need to be measured and mixed before use, hand laminated onto the repair and they take time to fully cure even when maintained at the correct temperature, making it time consuming and difficult to affect even a minor repair outside the workshop.

The Ultra V Patch, however, is a completely different product and is designed from the outset as an emergency repair system. It is compact, easily stored, ready for use straight from the pack and fully cures quickly using just sunlight. I know that similar products are used by canoeists to allow them to quickly repair their craft and continue their trip and I would imagine that the Connect Ultra V Patch would be ideal for track day and competition drivers wishing to quickly repair minor GRP damage on their cars so that they can continue with their activities rather than packing up and going home.

After only 10 minutes the repair was strong enough to return to service.

Bodywork & glass

Glass Etching

Martin Scott shows you how to make bespoke etchings on your kit car's glass.

For decades, mainstream manufacturers and main dealers have etched the registration number or VIN number on car windows as a security measure. If you're keeping your donor car's glass, the numbers won't match after your kit car has been issued with its own registration number by the DVLA, so what do you do? Maybe you get a bit creative, and have security and individuality at the same time!

My Quantum had the original Ford Fiesta registration number etched into the windows, so I thought I would try to modify it. I decided on a design that would cover the Fiesta's original registration number with a border and adding my own Quantum script using a DIY stencil.

I tried using expensive, specialist masking tape, but eventually settled on simply printing the script on a paper label! After that, peeling off the label, putting some clear tape on the wax backing paper, and re-applying the paper label gave me a durable stencil after I'd cut out the letters.

The width of the word is 67mm, so some delicate cutting was involved. The 'centres' of the Q and A were particularly fiddly, but not impossible to do.

As an option, getting some vinyl lettering made up would be a much easier and more accurate, if slightly more costly, approach.

Glass etching paste is available from art shops or similar outlets, and safety measures such as gloves and goggles need to be followed according to the manufacturer's guidelines.

Once the stencil is in place, the etching paste is applied with a small brush and left for about 30 minutes. After the residue is cleaned off, we have our personalised glazing! There's lots of scope here for kit car window design work, or car club glass tumblers with your own car featured!

If you give it a go yourself, send photos of your efforts into the magazine. We'd love to see what you create!

Original glass with donor car's now-obsolete registration number.

Paper label used to create the new design that will be etched into the glass.

Cutting out the stencil. A vinyl stencil would be an easier way to do this...

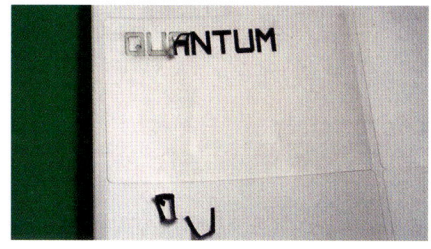
but paper is cheaper and, while more fiddly, not impossible to do.

Stencil in place, with additional clear tape border.

Etching paste at work.

The result. Give it a try yourself!

Modify, Improve & Upgrade Your Kit Car **39**

Fit An Aerocatch

If you've ever lost a bonnet on the move, you'll know the importance of a decent catch. **John Dickens** shows you how to fit the new AeroCatch 3 to your kit car.

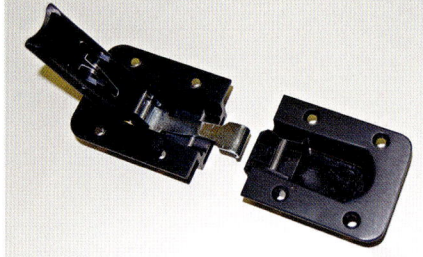

The new AeroCatch 3. The internal parts are clear when the catch is open.

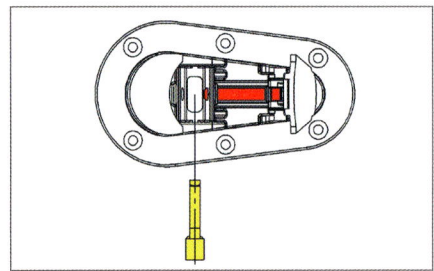

Original AeroCatch is basically a high-tech bonnet pin.

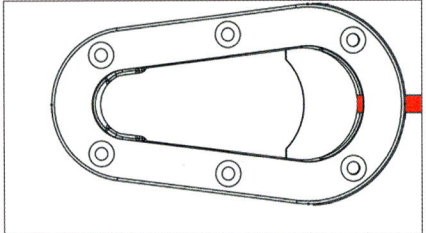

The AeroCatch 2 is similar but has longer sliding pin. We used them on our MEV Exocet (top).

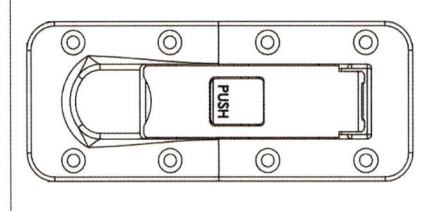

The AeroCatch 3 is a 2-piece over-centre type catch.

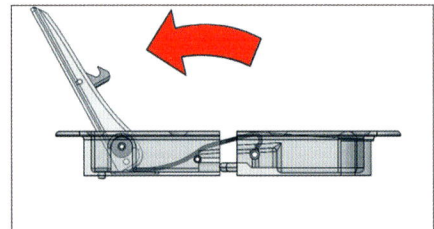

Internal catch is operated by an external locking lever.

This tongue aligns sections and carries shear loads.

Tongue locates in this slot as the catch engages.

Here the catch is closed but not yet engaged.

Bodywork & glass

Catch is now engaged and will lock when fully closed.

AeroCatch 3 as supplied. Don't lose the packaging.

Catches, fittings and instructions as supplied in the kit.

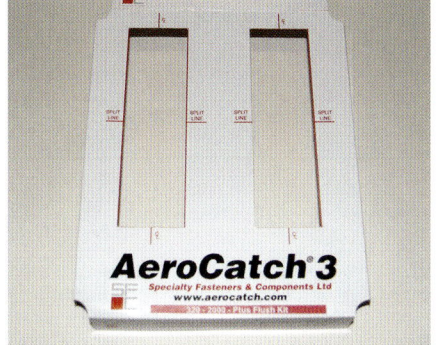
The packaging also contains the marking template.

These are the part panels John made for this guide.

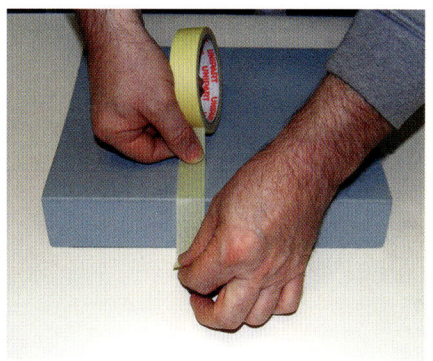
Protect the surrounding area with masking tape.

Masking tape is easier to mark than GRP or alloy.

Using square means catch will be mounted correctly.

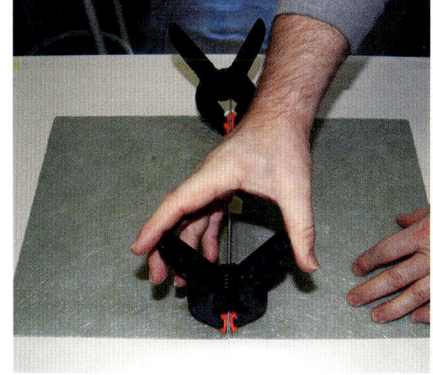
Two panels need to be clamped before marking.

The original AeroCatch has been with us for some time now and many of us are familiar with its operation. Essentially it is a flush fitting, lever operated, latching bonnet pin. The AeroCatch 2 was originally designed for marine use and is physically similar to the original version but the shear pin extends beyond the end of the catch to locate in a hole in the adjacent panel. The newest version, the AeroCatch 3, is a flush fitting version of the over-centre bonnet catch.

OPERATION
The AeroCatch 3 differs from the other versions in that it is a two-piece unit, although both sections are dimensionally identical in their fitment. A useful feature built into the catch is an engagement tongue which locates in a slot in the opposing section to align the panels as the catch engages. This tongue can also take shear loads in two directions, leaving only the panel clamping duties to the tension latch.

The catch is very positive in its operation. The two sections are aligned. They are located and pushed together then the lever is closed to engage the hook into the catch. As the lever is lowered it draws the two sections together and a positive click indicates that the secondary lock has engaged.

FITTING
The AeroCatch 3 is sold in a moulded plastic pack which clips together and can be opened without cutting or destroying it. The pack contains two complete catches, four load plates with captive nuts, sixteen countersunk Allen bolts and fitting instructions. The

Modify, Improve & Upgrade Your Kit Car **41**

catches themselves are black anodised aluminium with a brushed finish.

Once the catches are removed, the cardboard packaging contains a template for marking out the panel cut-outs, although the instructions also contain full dimensioned diagrams should you prefer this method.

Since I am not yet at this stage with my UVA, I moulded up a pair of small panel sections to demonstrate this fitting procedure. As always when working with soft materials such as GRP or aluminium, the first step is to tape up the area to prevent damage and to aid marking out. I used a square to ensure that the fixing centre line would be at 90deg to the panel split line. The two panels were then clamped together and the centre line was extended onto the second panel.

Since I was feeling my age particularly acutely this day I decided to go with the template to mark out the cutting lines. It has marked crosshairs which need to line up with the fastener centreline and the panel shut line. Ideally you would cut the template from the packaging and tape it to the panels but I needed to return the whole package intact so I simply held it in place whilst I marked the lines. These cut lines need to be extended over the panel returns so once again I used the square to keep these accurate and carefully measured and marked the cut depth.

There are various ways of cutting out the marked aperture. My own choice is a pad saw or mounted hacksaw blade. Power tools are too cumbersome for such small cuts. To start some cuts you may need to drill a hole

The catch centreline must match on both panels.

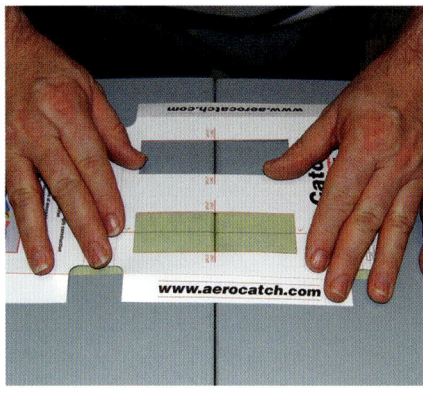
Template aligns with the centreline and shutline.

Should be trimmed and taped but needed returning.

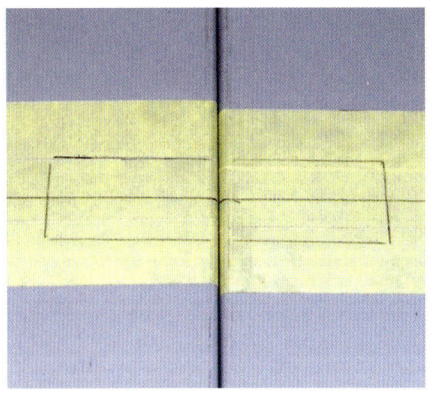
This is the first stage of the marking out.

The lines need to be extended onto panel return.

Cut depth here needs to be measured accurately.

This line completes the process.

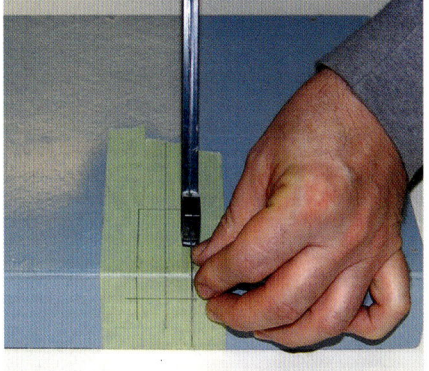
A pad saw or hacksaw blade with a handle is ideal.

Panels are now ready to have the catches fitted.

Bodywork & glass

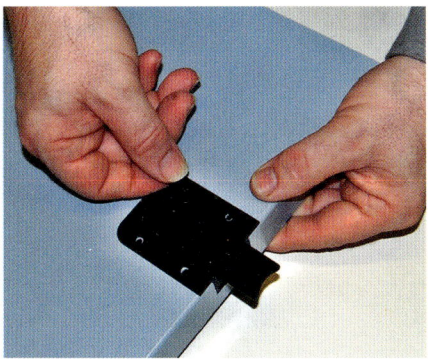
Position one catch and fit one fixing screw.

At the rear of the panel align one load plate.

The components are pretty much self aligning.

Gently tighten one screw to hold everything in place then fit the rest.

A couple of 5mm holes drilled on opposite corners will help.

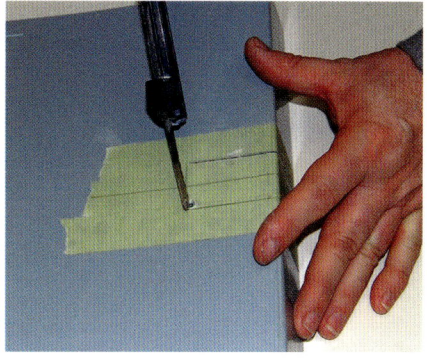
This straight cut is being started from one of the drilled holes.

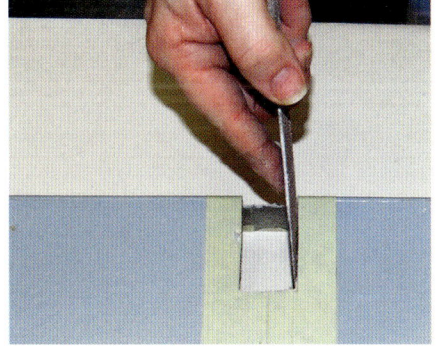
Always cut undersize then trim accurately using an abrasive tool.

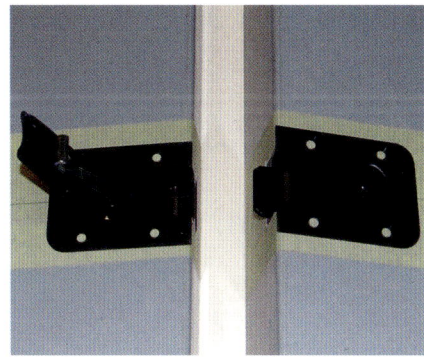
Check the fit frequently to avoid creating an oversized hole.

Using the square to align the catch as John marks the fixing holes.

Drilling the fixing holes at 4mm to allow for some wiggle room.

Use coarse abrasive paper to remove stray glass strands.

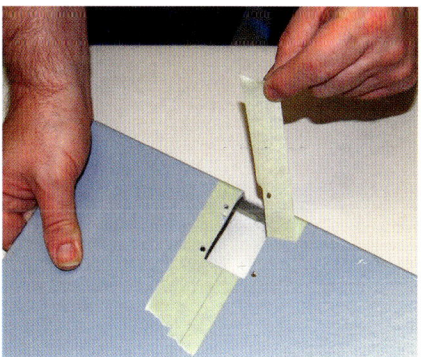
Finally remove the masking tape and wipe away any remaining dust.

Modify, Improve & Upgrade Your Kit Car **43**

Check fit is flush to panel before tightening screws.

These are the front and rear views of fitted catch.

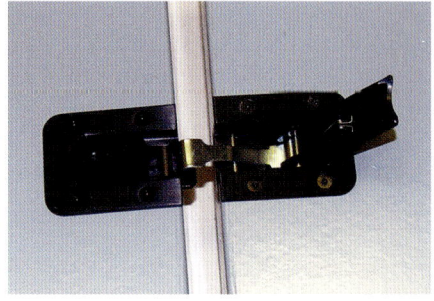
These are the panels in the open position.

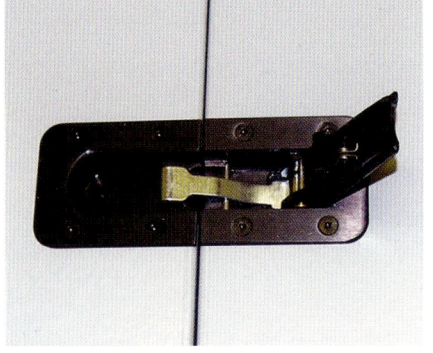
At this point, shear tongue has located the panels.

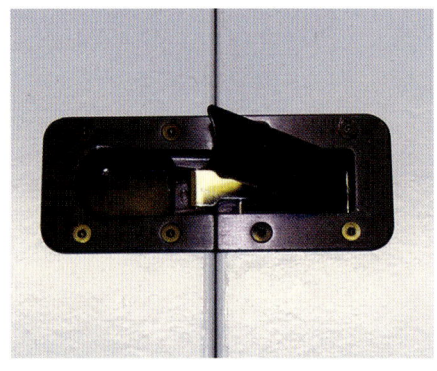
The catch engages as the lever is closed.

When lever closes fully, secondary lock engages.

This is the rear view of the closed panel.

Gaps allow for different panel thicknesses.

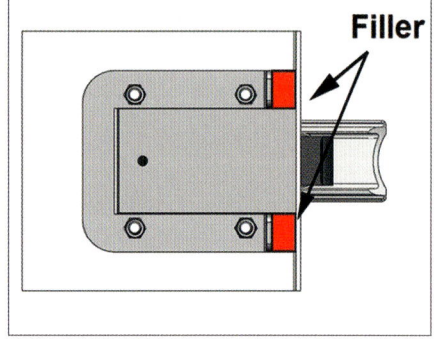
AeroCatch suggests that filler can be used for gaps.

at the corner of the waste material. As always, cut slightly inside the marked line so that final trimming can be done using a file, abrasive paper or a Permagrit tool. As you trim, keep checking the fit of the catch bodies.

If you are fitting the 325 Flush AeroCatches which fit from below the panel, this stage has to be done far more accurately as there is no upper flange to hide any gaps. The next step is to mark out the fixing holes. I used the square to make sure the catch was lined up correctly while I marked out the holes, then I drilled them using a 4mm drill. This gives a small amount of wiggle room with the 3mm bolts and aids final alignment.

If working with GRP, some rough abrasive paper can be used to clean up the edges of the cuts and holes. If working with aluminium you will need to use a file to de-burr these areas.

The last step before assembly is to remove the protective tape. To fit the catches, place one of the sections in its prepared aperture and fit one screw. At the rear of the panel use this screw to line up the load plate, then tighten the screw gently to secure the assembly. Fit the rest of the screws and tighten them gently. Check that this area of the catch sits flush with the panel return surface and fully tighten all the bolts.

The clamping action is very secure and the panel location is excellent. In order to accommodate different panel thicknesses and bend radii, the load plate stops short of the return face leaving a small gap. The manufacturer suggests that, for increased panel strength, you can fill the gap with a suitable, hard setting material.

At around £90 per pair these catches are not cheap to buy and they are trickier to fit than rubber bonnet hooks or over-centre bonnet catches, but the end result is much neater than either and, more importantly, far more secure too. The locating tongue prevents relative panel movement, so fretting damage is minimised and the secondary locking action of the catch means that accidental disengagement under load or vibration is impossible.

Finally, thanks to Europa Spares and AeroCatch for assisting with this article.

Bodywork & glass

Vinyl Wrapping

Increasingly popular and affordable, vinyl wrapping is an option more kit car owners are taking up – including **John Dickens**, who describes the process here.

Adhesive backed vinyl film has been used as an alternative to paint in vehicle graphics for many years now, but it is only relatively recently that it has been practical to cover an entire vehicle with a plain or digitally printed continuous vinyl film. This process is known as vinyl wrapping.

THE MATERIAL

The basic material for a vinyl wrapping film is polyvinyl chloride but other ingredients, such as plasticisers to make the film flexible, pigments to produce the desired colour, UV absorbers to improve resistance to UV radiation, heat stabilisers, fillers and processing aids, are also added to produce the required properties. There are two methods of producing the thin film (about 120 microns including adhesive) used in vehicle and other graphics.

Cast film is produced by pouring metered amounts of the liquid vinyl mix onto a constantly moving 'casting sheet'. This then passes through a series of heating processes which evaporate the solvents to leave a solid vinyl film.

Calandered film is produced by passing the vinyl mix, known as the 'melt', through a series of heated rollers until the required film thickness is obtained.

Not surprisingly, these two methods produce vinyl films with different properties and, due to its greater durability, conformability and stability, cast vinyl is preferred for vehicle wrapping.

Many companies produce films suitable for vehicle wrapping but the best known names are 3M and Avery. Development of these products is continuous and the latest vinyl films have an 'air release' feature which helps to prevent air bubbles. The adhesive side of the vinyl has continuous air channels which allow any trapped air to be removed during application. Some films also have glass beads in the adhesive layer which prevents it from bonding firmly until pressure is applied. This allows the film to

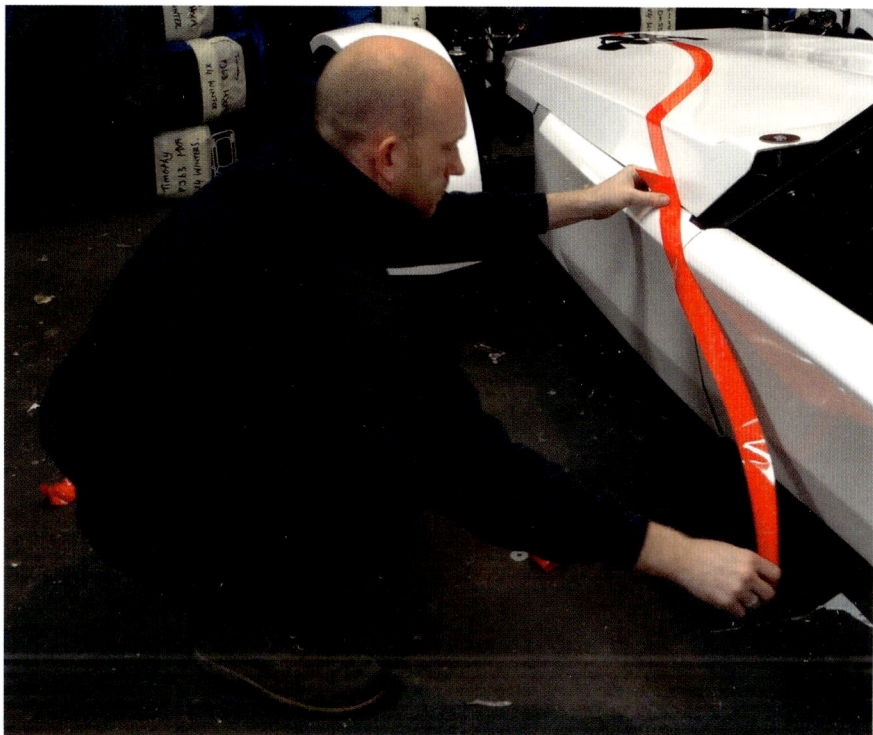

be repositioned a number of times without damage or distortion.

THE WRAPPING PROCESS

It had always been my intention to have my UVA Fugitive wrapped rather than painted and, although I had been assured by a number of people that it was possible to do the work myself, I had always intended to have it done professionally too. When the car was finally finished, I looked around for a company to do the work. Perhaps I was being naïve, but I imagined that this would be a relatively straightforward process. In fact, of the sixteen companies I originally contacted only two were happy to do the work while allowing me to be present to take notes and photographs. Only one, the company featured here, was professional enough in their preparation and performance to give me the confidence to use them. The rest either ignored me completely, agreed to do the job and then ignored me or wanted to work with their own designs rather than mine.

The company I chose was Premier Graphics and Signs in Durham. The owner, Paul Wadge, has been working with vinyls for over twenty years and has been using them to wrap vehicles since the process began to gain in popularity around ten years ago. Paul initially came to my home to see the car, take measurements and photographs and to talk to me about my proposed design and what was possible with the materials available. In spite of all his experience, Paul had never wrapped directly on to a GRP gelcoat surface before. Sensibly, therefore, before he was prepared to take on the job he wanted to check the adhesion of the chosen vinyl onto the gelcoat surface and he returned a few days later with some samples to test.

In the meantime, I had cleaned and polished all the UVA panels with Farecla G3 to remove any chalking and oxidation as

John chose Premier after contacting many firms.

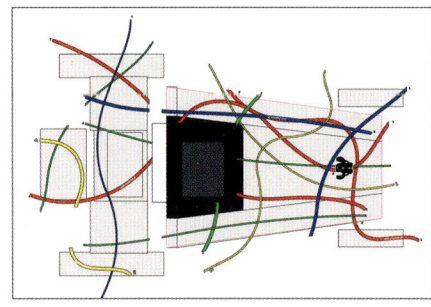
Scale plan of car was used to design the graphics.

Lights, fittings and rear mudguards were removed.

Vinyl sheet is cut to size with plenty of overlap.

The panel is washed with water then degreased.

With backing removed, sheet is pressed into place.

Felt edged squeegee used to avoid marking vinyl.

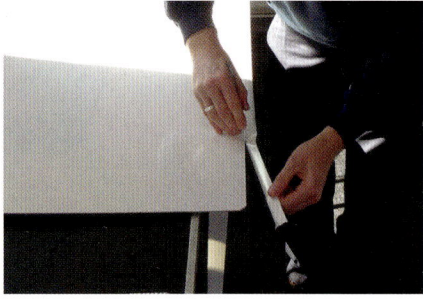
Scalpel used to trim the vinyl flush to the edge.

Paul also cut the vinyl away from all the fixing holes.

some of the panels were four or five years old by now. After trying a couple of vinyl samples on the panels, Paul was satisfied that the wrap would stick properly and was happy to proceed. The next step was to meet with Tom the designer, at the company's premises, to discuss and finalise the design so that the materials could be ordered and prepared. I had already sent them a photograph of a car with a design I quite liked (a classic Opel touring car) and Tom had created an accurate flat plan of the car with a design based around this colour scheme.

Initially he had produced a symmetrical design with the majority of the stripes running along the length of the car, but when he realised that I wanted a random arrangement of stripes over the entire car he took a different approach. Rather than striping each panel individually he laid curved stripes over the whole car. My original idea was to have the stripes digitally printed onto white vinyl before wrapping it on to the car but Paul advised against this as, once he began to stretch the vinyl around the more complex curves, the stripes would distort or bend.

The revised plan was to put on a complete single colour wrap then add the stripes and logos afterwards. He also suggested that I might like to look at some of the alternatives to a plain white base colour and showed me a selection of gloss, satin, matt, metallic and pearl finish white vinyl materials. Faced with such a choice, I abandoned my original idea of gloss white and instead chose a pearl white vinyl which has a gold sheen in bright light.

With all the design decisions made, we arranged a date to actually wrap the car. I took the UVA down to the workshop the afternoon before we were due to start and stripped off all the lights, locks and rear mudguards. I also removed the bonnet and engine covers for ease of access. All the loose panels were then taken to the office workspace where the light and temperature were ideal.

Paul started with a small flat panel, the engine cover. The vinyl was trimmed roughly to size then the panel surface was cleaned with water then degreased with panel wipe. The backing paper was removed from the vinyl and it was carefully laid onto the panel. Paul then began pressing the vinyl onto the panel using a felt edged squeegee. The soft felt edge prevents the squeegee marking the vinyl surface. Once Paul was happy that the vinyl had stuck properly with no air bubbles he used a scalpel to trim off the excess vinyl.

On a production car, the vinyl would be wrapped over the panel edges onto the returns but on my car there are no returns so the vinyl was simply trimmed flush with the edge of the panel. He also used the scalpel to carefully cut out all the fixing holes so

Bodywork & glass

Gentle heat allowed vinyl to stretch around curves.

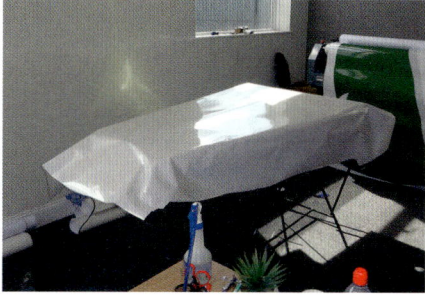
A single sheet of vinyl was lowered onto the bonnet.

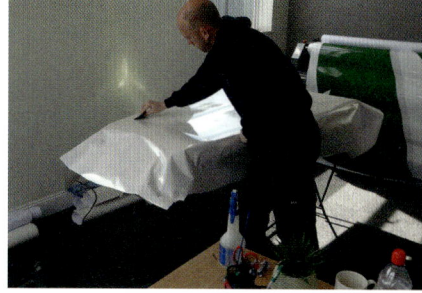
Starting at the centre, Paul worked vinyl into place.

Vinyl can be lifted and repositioned for the best fit.

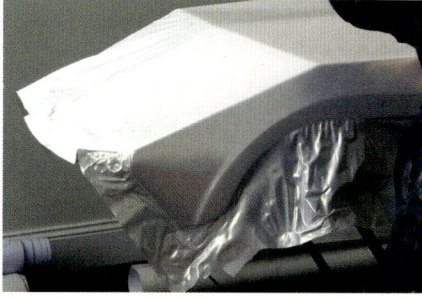
Heating the vinyl allowed Paul to get a perfect fit.

Scalpel blades changed regularly for accurate cuts.

John chose a textured carbon fibre wrap for roof.

These panels are bigger than they look; quite tricky.

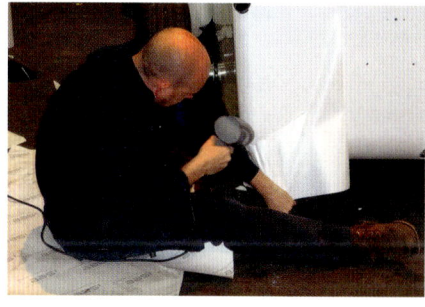
Smaller section was needed where curves meet.

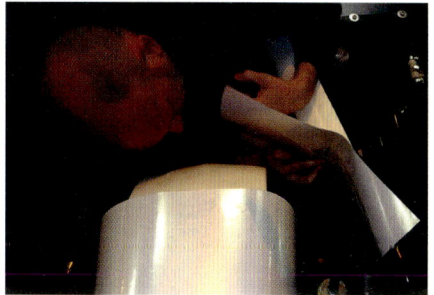
Complex curve wrapped with separate vinyl patch.

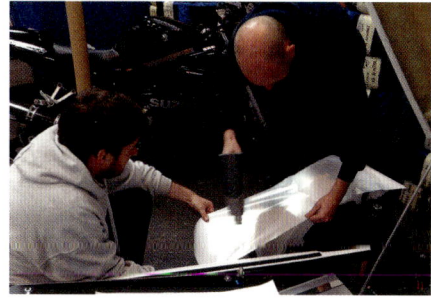
Main mudguard face and sides wrapped separately.

Printer used to print digital images on vinyl sheets.

There is no limit to what can be printed!

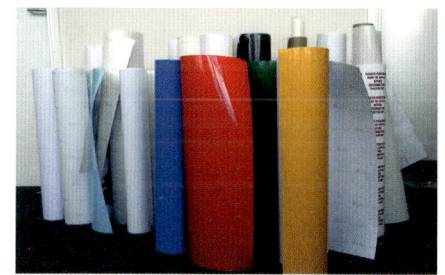
Self-coloured vinyls were used for the stripes on UVA.

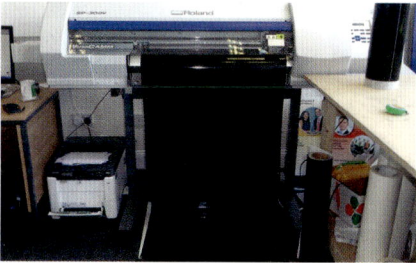
This machine cuts the graphics from the vinyl sheet.

Modify, Improve & Upgrade Your Kit Car **47**

that the vinyl would not lift as the fasteners were pushed through during reassembly. The scalpel blade was changed frequently to be sure of a clean cut every time. On flat surfaces or single curves heat is not needed but for more complex curves Paul carefully used a hot air gun to soften the vinyl so that it could be shrunk or stretched as required. The vinyl can be lifted and repositioned a number of times to ensure a perfect fit but this particular vinyl had a maximum stretch of 30 per cent and Paul was careful not to exceed this. If the vinyl is overstretched when fitting, it can slowly shrink back leaving voids.

The bonnet was wrapped next. This is the largest panel on the car and I imagined that Paul would probably peel off the backing paper a small section at a time as he smoothed the vinyl down. In fact, he and his assistant Ian simply peeled off all the backing paper at once and, holding the corners of the vinyl, laid the whole sheet onto the bonnet. He then began to carefully squeegee out all the air, lifting and repositioning the vinyl a number of times before he was happy. Once again, heat was needed around the more complex sections but the final fit was quite remarkable. The rear mudguards are curved in two planes and have a tight radius at the edge, so Paul's initial feeling was that he may have to wrap the vertical sides separately to avoid wrinkles but, once he started working and warming the vinyl, he found that the whole mudguard could be wrapped with a single vinyl sheet. Once again the excess was carefully trimmed away.

When all the removable panels were wrapped, we moved back to the workshop to continue the process on the rest of the car. My plan was that the roof panel should effectively 'disappear' into the chassis frame so that was wrapped in textured carbon fibre. The engine cover sides were done next so that I could refit the mudguards while Paul worked on the rest of the panels. They are larger than they look and have some tricky contours on them so, even with careful heating, it was obvious that they could not be done in one piece. Small patches were used and the seams were hidden out of sight under the car. A similar process was used in the tight corners at the front and rear of the side panels. The seams are hidden on the very edges of the contour changes and are invisible unless you know exactly where to look.

Although the front mudguards look similar to the rears, they are smaller and the edge radius is tighter so Paul did use separate pieces for the inner and outer rims to eliminate the possibility of wrinkles. Once again, the seams are all but invisible. By the time Paul had finished wrapping the remaining panels, I had the rear mudguards in place.

Once the white base wrap was complete the stripes could be applied. Had I chosen a digitally printed design, the pattern, image or graphic would be digitally printed onto the vinyl sheet then a thin clear vinyl film would be heat laminated over the print to protect it from physical damage and also to minimise the effects of UV light. Depending on the amount of exposure the vehicle gets, digital prints have a life of around seven to eight years. The stripes used on my design were cut from self-coloured vinyl so they have a much longer working life. The machine that cuts the graphics is controlled by the computer used to compile the original design and uses a stylus which can accurately cut through the vinyl layer but leave the backing paper untouched. The waste material is removed leaving the graphics attached to the backing sheet.

The resulting coloured stripes were curved to match the original design, sectioned to allow for panel joins and numbered so that they could be indexed to the design drawing. As Paul applied the stripes we made some minor alterations to the pattern for ease of fitting but the final result, especially over the bonnet area, was essentially just as it was designed.

Wherever the stripes crossed, Paul cut away the upper stripe producing a very neat effect (Fig 35). The small 'UVA Fugitive II' logos, some curved and some straight, were cut from white vinyl and added on top of the stripes but another option would have been to cut the logos out of the stripes so that the pearl white wrap would show through.

The only way to enter and leave my car is to slide over the side panels so, to avoid damaging the graphics, a layer of clear 'stone chip protection' vinyl was added here and on the rear wheel arches. This was the only time in the whole process that Paul applied the vinyl onto a wet surface to allow more working time and flexibility.

With the wrap completed, I refitted all the lights and latches to complete the rebuild. As always, the refitting of the exterior trim really

Waste vinyl removed, leaving the cut graphics.

Each stripe shaped and indexed according to design.

Stripes were cut around the logo and crossing points.

These smaller logos are applied onto the stripes.

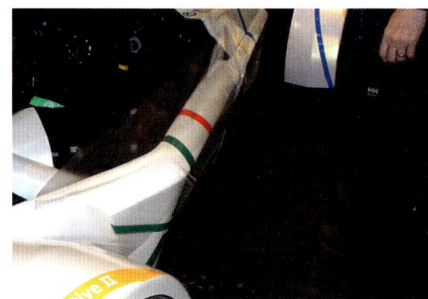
Clear 'Stone Chip' vinyl applied to high wear areas.

Bodywork & glass

Inspiration came from an Opel touring car.

Refitting all the exterior trim really brings out the finish.

brought the finish to life and the final result exceeded all my expectations. Personally I think it looks stunning. The cost for the whole process was a very reasonable £845 plus VAT which brought the final total to £1014. I would imagine that an equivalent pearl white paint finish with graphics would cost double this and, in addition to the exterior trim removal, would also involve preparation, masking, drying, curing and polishing. Since I was photographing the process for this article, Premier Graphics cleared their schedule and made a special effort to complete the whole wrap in one working day but two or three days would be typical for a full vehicle wrap.

WRAP OR PAINT?

Obviously both finishing systems have their pros and cons and I have outlined the main ones here:

Preparation. There is very little difference in the surface preparation required for painting or wrapping. The vinyl film is actually thinner than a combination of primer and paint so any surface imperfections must be corrected before applying the vinyl. If the vehicle is already painted the finish must be sound. Poor paint finishes may be damaged as the vinyl is lifted and reapplied during the wrapping process.

Cost. The cost of a vinyl wrap will depend on the design chosen. A digitally printed wrap will cost more than a single self-coloured vinyl for example but in general wrapping is cheaper than a full respray.

Options. Vinyl films are available in a huge range of colours and in gloss, satin, matt, candy, pearl and true metallic finishes. In addition any design you want can be digitally printed onto the vinyl before wrapping. Paint finishes cannot offer this range of options.

Time. Some aspects of preparation, such as the removal of the trim and lights, are the same for a wrap or a paint job, but there are some major differences too. With a vinyl wrap there is no masking off, no drying, curing or baking time and no need to buff or polish the final finish. As soon as the wrap is applied the car can be driven away. Typically a full wrap takes two to three days.

Durability. Vinyl films are improving in this respect but they are still not as hard or durable as a two-pack paint finish. Exposure to sunlight (UV) will cause the colours to fade but even digital prints should last seven to eight years in normal use.

Protection. A vinyl wrap can be used to protect an original paint finish from degradation and the film can be removed to reveal the original finish if desired.

Removal. Vinyl films can be removed and replaced far more easily than respraying the vehicle provided the underlying finish is sound.

IN CONCLUSION

In the past, I have always painted my own cars but, over time, this has become more problematic, first with the introduction of two-pack polyurethanes then later with water based coatings. As a result, I was curious to see the wrapping process first hand and, although this has been my first experience of the process, I have to say that I am already a convert. I cannot imagine myself ever painting a car again. The range of finishes available, the time saving and the cost means that for me the choice of a vinyl wrap is a no-brainer.

The roof panel effectively 'disappears' with the carbon wrap.

Powdercoating

John Dickens shows you how to use Electrostatic Magic's latest DIY powdercoating system, which claims to be easier than ever.

This is the original system with the power pack and foot switch.

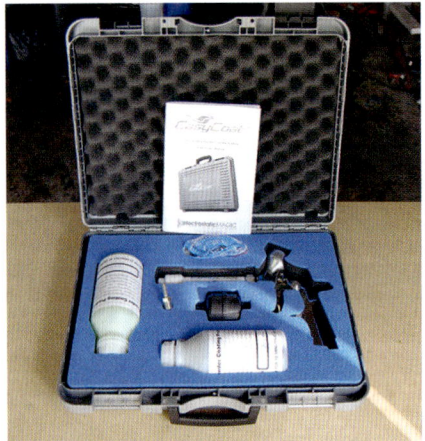

The new system, as supplied in the basic kit, is much simpler.

An electric oven is ideal to cure small items. John's is a secomd-hand Smeg.

Some time ago I wrote an article comparing different methods of surface coating designed to protect steel components from corrosion. I looked in detail at three particular methods. Hammerite (a hammer-finish paint), Rustseal (a moisture cured polyurethane paint) and powdercoating (a heat cured finish applied electrostatically as a dry powder).

Until relatively recently, the successful application of a powdercoating had been an industry-only process requiring specialist equipment to apply and cure the coating. In the last few years, however, a number of companies have been offering small scale equipment for DIY use.

The kit I used for the original article was supplied by Electrostatic Magic and, in fact, I have been using one ever since on small components for my own car and motorcycle. The company regularly attends kit car shows and, for my own personal interest,

Rust-proofing

This particular rust convertor is ideal for high temperature use.

U-Pol acid etch primer is temperature stable and perfect for alloy.

Different sized containers are available for large and small jobs.

This shows the gun assembled and ready for use.

John's small compressor is more than adequate to power the spray gun.

This adjustment controls the output pressure of the compressor.

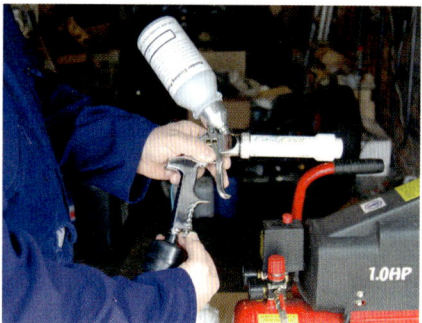
The regulator screw is closed completely before setting the pressure.

I always visit its stand to see if it has any new products which I may find useful. At Stoneleigh this year Electrostatic Magic was demonstrating its latest EasyCoat powdercoating system which has been completely redesigned to make it much simpler and far more convenient to use than the original product.

The old system used an electrical power unit to positively charge the powder as it passed along the spray gun and to negatively charge the item being coated so that the powder would be attracted and stick until cured. The unit was controlled by an external foot switch which activated the charging system as the powder was applied. Whilst this was not difficult to set up, the time taken sometimes meant that, for one or two small objects, I might opt to use an aerosol paint can instead.

The new system is far more convenient as it does away with the electrical charging system entirely. The internal structure of the spray gun positively charges the powder by friction as it passes along the barrel and the object being coated is negatively charged by being connected to the body of the gun. The powder coats the object and sticks due to the attraction between the opposite charges.

As before, a compressor is needed to fluidise the powder but once this is connected and set up, you simply put powder in the cup, attach it to the gun and you are ready to spray.

The basic kit costs £179 and contains...
- Powdercoating gun
- Two powder canisters
- Moisture separator/filter
- Ground strap with wristband
- 1/4 BSP universal barbed hose fitting
- Custom foam lined plastic moulded case
- 1/2 litre of powder (choice of colours)
- Instruction booklet

The instruction booklet supplied with the kit outlines the process itself, details the health and safety aspects, and explains how to set up the equipment and how to use it to coat metals, glass, ceramics and wood. There is also a troubleshooting section should any problems arise.

The starter pack, which contains all of the above plus more powders, safety equipment, spare cups and filter, rust convertor and masking tape, costs £239.

To melt and cure the powder it must be heated at 180deg C for 10 minutes. This can be done in a normal domestic electric oven but there are other methods too. Don't use the oven in your kitchen, as fumes are produced as the powder cures. I have an old Smeg oven in my utility room which I bought on eBay.

IN USE
To demonstrate the process I used the equipment to coat a steel bracket I had made to mount my UVA handbrake lever. The first step is to prepare the object for

Modify, Improve & Upgrade Your Kit Car

powdercoating. This is very similar to preparation for painting. All traces of previous coatings, rust, oil or grease must be removed. Rust can 'outgas' when heated, producing a pitted surface and poor adhesion.

The company recommends and sells a rust convertor called Brunox which converts rust to black iron tannate and is stable at the curing temperature. The powdercoating can successfully be applied directly to clean metal, but should you prefer to use a primer, especially on aluminium, the company recommends U-Pol Acid 8 etch primer which is also suitable for high temperature use.

I had prepared my bracket using electrolysis to remove the rust, followed by an acid wash to etch the surface, so I decided against using primer. The chosen powder colour needs to be poured into one of the containers. There is a huge choice of powders including solid colours, metallics, candies, flakes and clear coatings but I tend to favour black for most components. The powder supplied was satin black which was even better.

The filled container is then screwed onto the gun, the hose adaptor is screwed into the filter unit and this assembly is screwed onto the bottom of the spray gun. The earth lead is clipped onto the earthing post and the gun is ready for use.

You do need a suitable compressor to operate the gun. I have a small 1.0hp unit which produces about 4cfm and has a 6-litre air receiver. It cost about £75 when I bought it a couple of years ago.

Once the output hose is connected to the hose adaptor on the gun, the output pressure needs to be set on the compressor. The equipment has a maximum working pressure of 50psi but may not need this pressure to operate correctly. Close the regulator screw on the gun fully then press the trigger completely and adjust the air pressure until a gradual flow of powder comes out of the nozzle. On my gun this happened at around 40psi. Once the pressure is set the gun is ready for use.

I made up a wire frame to support the bracket then degreased it using acetone but thinners or panel wipe would be fine too. If anything needs masking off it should be done at this stage. High temperature masking tape is available, but you can use ordinary tape if you remove it while the object is still warm.

Since the whole point of the process is to fluidise and disperse fine powder, Electrostatic Magic recommends and supplies a dust mask and goggles rather than safety glasses.

Finally, the earth lead is clipped onto the bracket and the regulator screw is opened a small amount so that when the trigger is pressed a small cloud of powder sprays from the nozzle. The powder should be applied from about 5in away from the object. It will be attracted to the item and stick to it. Rotate the object to ensure full coverage and take extra care with internal corners.

It is impossible to apply too much powder as the coating insulates the metal and prevents any excess powder sticking. There will therefore be some waste during the coating process as not all the powder sprayed by the gun will stick to the object. This can be re-used if it is collected on a clean plastic sheet or in a clean dry container. It should be filtered through a 400 micron filter (such as a paint strainer or pool filter bag) to remove debris before re-use.

Once the object is fully covered, carefully remove the earth lead and spray the region where the clip was attached to cover the small area of bare metal left behind. Carefully, without

The pressure is adjusted until powder just flows from the nozzle.

John made this hanger from welding rod. Coat hangers are good too.

The bracket is degreased with acetone (since that's what John had handy).

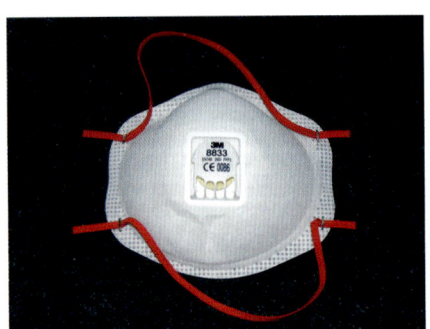

A good dust mask should be worn to avoid inhaling the powder.

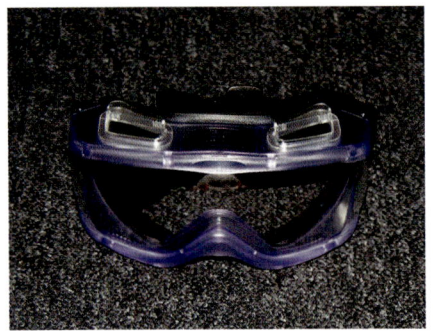

Goggles are far more effective than safety glasses in this application.

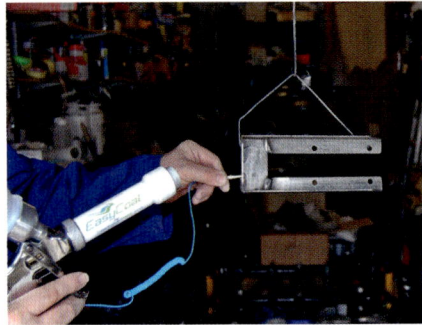

The earth lead needs to make a good connection to clean metal.

Rust-proofing

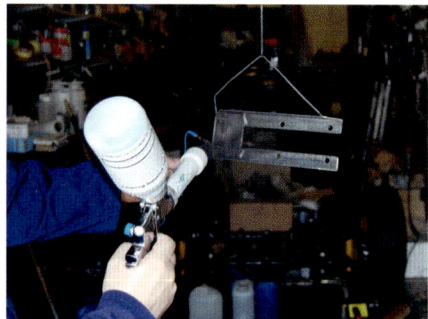
Use a gentle cloud of powder about 5in away from the item.

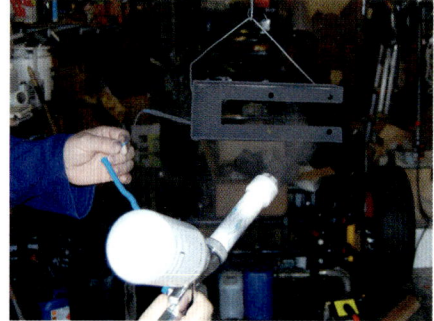
The powder will slowly build up on the component to form a complete layer.

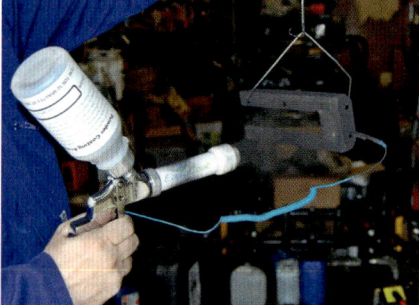
Turn the item round to spray all the surfaces. Check for gaps.

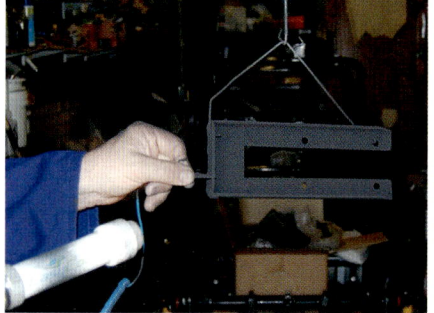
Carefully disconnect the earth lead when coverage is complete.

Remember to spray the area left bare by the earth lead clip.

Carefully transfer the object to the oven and hang it from a shelf.

The coating has a matt appearance before curing.

Start timing 10 mins when all powder has melted.

Finished coating is ready as soon as it's cool.

touching the coating, transfer the object to the oven. It will have an overall matt finish prior to curing.

Set the oven to 180deg C and switch on. The powder takes 10 minutes to cure once the object has reached the correct temperature. Keep an eye on the powder. As it melts it will turn glossy. When all of the powder has melted begin timing and allow a full 10 minutes. When the time is up, turn off the oven and open the door slightly to allow the heat to disperse slowly. The coating can crack if cooled too quickly.

If you have used any masking tape remove it whilst the object is still warm. Once the item is cool it is ready for use. There is no further drying or curing time. After use, or to change colour, the gun is cleaned out using compressed air to blow out the powder. There are detailed instructions in the booklet.

Obviously, larger objects need a larger oven to cure the coating. At the Stoneleigh show Carl, from Electrostatic Magic, was explaining how one customer had made an oven using a galvanised steel dustbin and a halogen cooker, and has subsequently posted details of this on the company website. He also sent me one of the photographs of this oven. Alternatively the company suggests that larger objects can be cured using an infra-red lamp or a hot-air paint stripping gun.

One customer made this oven from a dustbin and a halogen cooker.

Modify, Improve & Upgrade Your Kit Car

Zinc Plating

Zinc plating is now within the reach of DIY enthusiasts thanks to readily available kits. **Ian Stent** joins ACKCC member Trevor Rolph to try one out.

You must get the component to be plated as clean and rustfree as possible.

The plating kit comes packed in a 5-litre bucket.

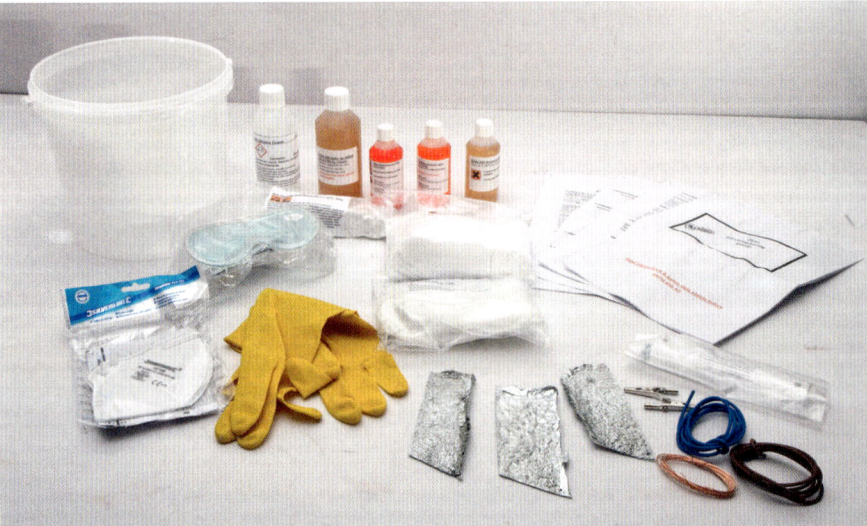

The complete kit laid out.

Rust-proofing

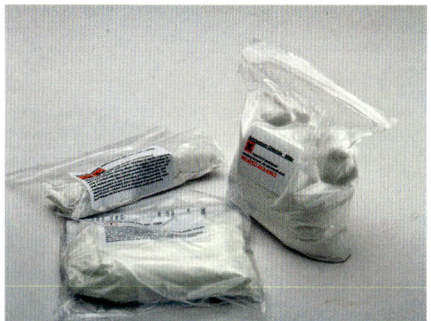
Dry acid salts, ammonium chloride, zinc chloride.

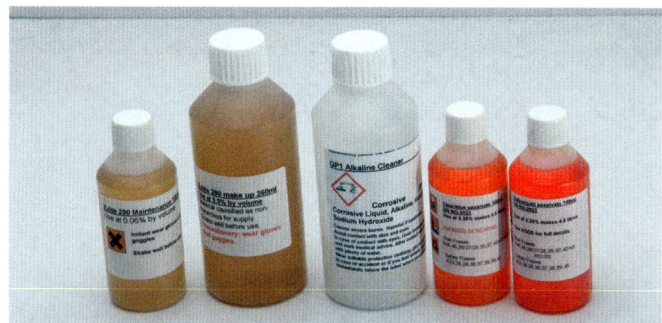
Cleaners, passivates and other liquids.

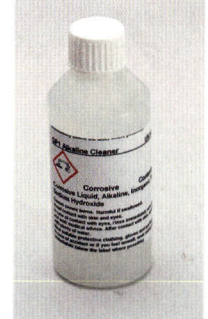
Alkaline cleaner.

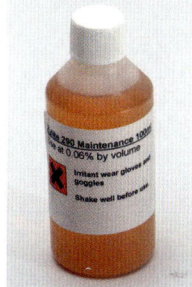
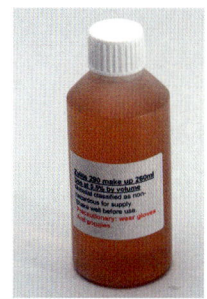
Maintenance fluid should be shaken before adding to the plating solution. Make up fluid needs adding.

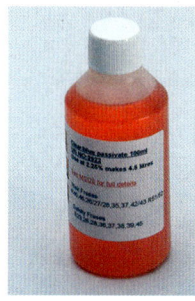
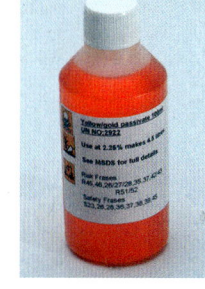
Passivates typically come in yellow or blue and will give the item an attractive OEM hue.

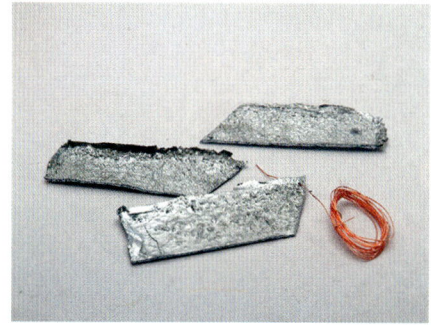
Zinc anodes.

Kit car builders are always on the look out for ways to stop metal components from corroding while, if possible, improving the visual appearance of a part at the same time. This is usually achieved by coating the metal with a covering of some description... typically paint or powdercoating. Paint has been the easiest solution for most of us, requiring little or no specialist knowledge and no specialist tools. And powdercoating is now also within the reach of the home enthusiast thanks to kits such as that supplied by Electrostatic Magic.

Beyond the reach of the amateur enthusiast is galvanising, a process of dipping components (usually a complete chassis) into a molten bath of zinc. While it gives excellent rust protection, the finish has an industrial look that's hard to improve with conventional paints. What's more, it's far from ideal for smaller, more delicate components since the coating is comparatively thick and heat distortion can be a risk.

But there is another option which is ideally suited to brackets and smaller items, and which can give a visually appealing OEM quality of finish... plating.

Zinc plating coats materials in an ultra thin layer of zinc, most usually achieved by a process of electroplating within a solution bath. Various kits are now available to the amateur and are perfectly suited to smaller components such as engine ancillaries and brackets. Once coated, these materials will have a bright and shiny finish and can be further polished. Vitally, they also achieve the main objective of offering a good surface resistance to corrosion.

Zinc plating can be used on a variety of materials, such as steel, iron, brass and copper.

PLATING KITS

A quick search on Google or eBay will reveal various plating kits, typically starting from around £50. The kit will contain a number of different components (powders and liquids) and these can be ordered individually at a later stage should you run out. The mixed electrolyte liquid itself is said to last years if kept clean, so the product potentially has a long shelf life and can be used repeatedly.

WHAT'S IN THE BOX

The kit we used all arrived inside the 5-litre bucket in which the process could be done. Contents included...
- Safety gear – Goggles, inspection gloves and a basic dust mask.
- Current controller
- Electrolyte salts
- Ammonium chloride
- Zinc chloride
- Passivates
- Cable and crocodile clips
- Make up solution
- Maintenance solution
- Zinc anodes
- Copper wire
- Set of instructions

OTHER ITEMS

In addition to the kit itself, you'll need some other items. These are...
- De-ionised water – Four litres of de-ionised water are needed. You can get this from an autofactor.
- Battery charger – Needed to provide the power for the electrolysis to work. One with an ammeter can he helpful.
- Air pump – Not vital, but the solution needs to be agitated throughout the plating process and this is best done with a pump, such as those used in fish tanks.
- Measuring jug
- Copper pipe

SAFETY FIRST

Some of the chemicals used in the process are corrosive, so it's very important to wear the supplied gloves, goggles and face mask when required. The process itself should also be done in a well ventilated area, such as your garage with the door open.

Modify, Improve & Upgrade Your Kit Car

BEFORE YOU START

Before you can start plating, there are various jobs that must be done first. Having removed everything from the bucket, the first job is to attach a short length of the bare copper wire to the zinc anodes. A small hole had been pre-drilled into the anodes for this purpose, so the wire was threaded through the hole and then twisted back on itself.

The three anodes are then hung inside the bucket, using the wire as a hook over the top lip. More of the remaining copper wire is then wound around the outside of the bucket and the three wires from the anodes are twisted onto it to form a good connection.

The plating solution can now be made. Our pack recommended pouring in four litres of the de-ionised water. Ideally, this should be warm. One bag of ammonium chloride is then added in its entirety, along with one of the supplied bags of zinc chloride. These should be stirred well until everything is dissolved.

The bottle of make-up solution needs shaking before the specified amount is added to the bucket (the rate is given in the instructions, in this instance 5.5 per cent by volume). Then the maintenance solution is added (0.06 per cent by volume). The plating solution is now ready to use.

Three other solutions can also be made at this stage, the alkaline cleaner, dry acid salts solution (pickling) and passivate solution. The alkaline cleaner is supplied in a liquid form and just needs pouring into a container to allow the item to be submerged prior to plating. The dry acid salts need dissolving in water at the specified rate, once again in a volume which allows the item to be fully submerged immediately prior to plating. Passivate is used after the plating process. Different colours are available and the passivate liquid is added to water at a specified volume.

The current controller can also be assembled at this stage. It consists of a half-round section of plastic tube with a hole at each end. Into these are inserted two M6 nuts and bolts and a wound wire is connected across the two bolts.

A length of the sheathed cable then needs attaching to the supplied crocodile clips so that at one end it can be clipped onto one end of the copper wire wound around the bucket and, at the other end, attached to one end of the current controller.

The positive output from your battery charger is then attached to the other end of the current controller.

The copper pipe needs bending so that it can rest across the bucket and items to be plated can be hung from it. The negative connection from the battery charger is attached to the copper pipe, but take care that the pipe does not touch the exposed copper wire wound around the bucket. You are now ready to have a go.

PROCESSES

There are various processes involved in plating a component and, in order, these are...
- Component preparation
- Alkaline cleaning solution
- Rinse in water
- Pickling
- Rinse in water
- Plating
- Rinse in water
- Passivate
- Rinse in water
- Enjoy!

THE TEST

The plating kit we used was bought by Apple County Kitcar Club member Trevor Rolph and he was hoping to use it to plate various components on the Honda MT5 motorbike he's been restoring.

Clearly, you need to ensure the part you are trying to plate is as clean as possible. It should also be free from grease and any

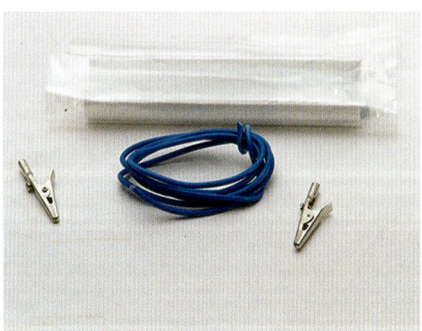
Current controller unit with cable and clips.

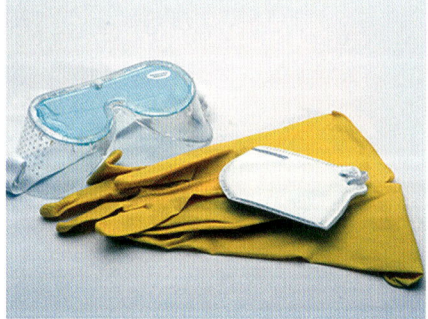
Safety kit is supplied for a reason... use it!

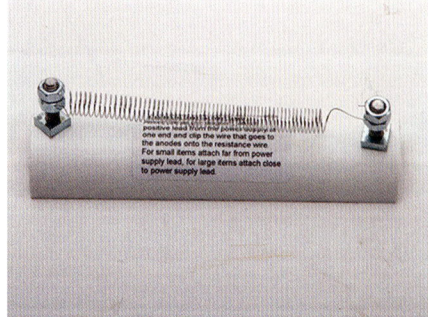
The current controller now assembled.

Preparing the zinc anodes with the copper wire.

Zinc anodes are hung evenly around the bucket using the copper wire as hooks.

More copper wire is wound around outside of bucket making contact with wire from the three zinc anodes.

Rust-proofing

PVC wrapped electrical cable with crocodile clips attached...

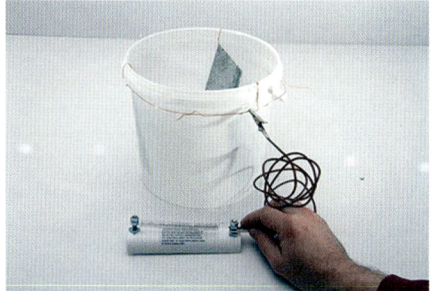

...One end attached to the wire and one to the current controller.

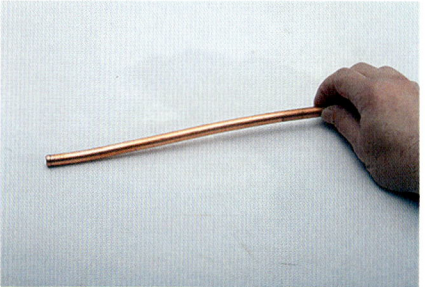

You'll need to source some copper pipe to act as a hanger for the items to be plated...

...It needs bending, but must not make contact with the copper wire around the bucket.

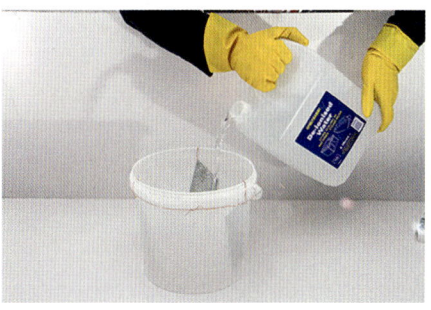

De-ionised water being poured into the bucket.

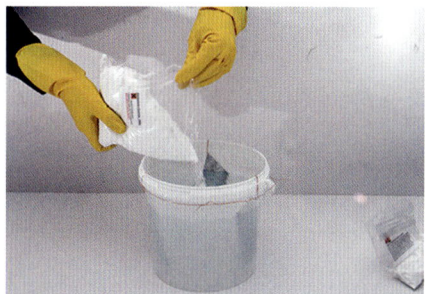

Pouring in a complete bag of ammonium chloride, and then one of zinc chloride.

Pouring in the maintenance fluid, followed by...

...a specified number of capfuls of make-up fluid.

The solution needs to be thoroughly stirred

other deposits. As with so many kit car jobs, the better the preparation, the better the end result. Once the item is as clean and rust free as you can get it, you are ready to begin the process.

It has to be said that between Trevor and the CKC team, we found measuring out the various liquids in the right quantity to be quite a challenge! For starters, we did not have an accurate measuring jug and I think a syringe would have been useful to measure the smaller quantities. But we did get there in the end, and assembling the rest of the components was easy. Trevor had brought along his battery charger which had a meter reading on it, which proved handy for adjusting the current controller.

Agitating the plating solution is recommended and using some form of air pump is suggested. However, the airbed pump Trevor brought with him was far too powerful and, even after typical kit car ingenuity, was far from ideal. Since our test he's bought a fish tank pump (available on eBay for under £10) which delivers a much more controlled level of water agitation.

In our haste to try out the process, we inadvertently missed the instructions regarding the alkaline cleaning process and subsequent pickling process. As such, the results achieved were impressive, but would have been better had we done these two pre-plating processes... the bracket should have been left in the alkaline solution for around five minutes, before being rinsed with water and then immersed in the pickling solution for a further minute. A final rinse in water and then into the plating solution, taking care not to touch it with your hands.

The small bracket Trevor wanted to plate was attached to a length of copper wire and suspended in the plating solution. The battery charger could now be switched on and the water pump engaged. The plating process begins almost immediately and, from the online guides we'd watched, the process should be visible on the component within as little as 60 seconds. Our test didn't yield such instantaneous results (quite possibly because of our lack of preparation!), but after the recommended 30 minutes we felt we had achieved a pretty good result given that it was our first attempt.

The final job was to switch off the battery charger, remove the item, rinse it under water and put it into the passivate solution for around 30 seconds. The passivate is typically available in either clear or

Modify, Improve & Upgrade Your Kit Car

Make sure the copper wire holding the zinc anodes is out of the plating solution throughout the process.

We first tried an air bed pump to agitate the solution, but a fish tank pump is much more suitable.

After being treated in the alkaline cleaner and pickling solution, the item to be plated can be lowered into the plating bucket.

The cable from the current controller goes to the outer copper wire...

...the battery charger's earth lead goes to the copper pipe supporting the item to be treated...

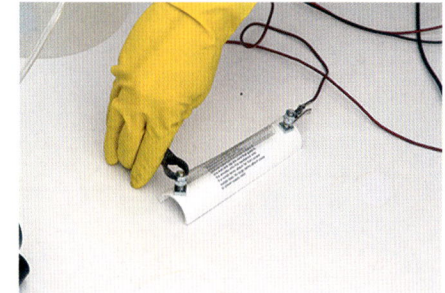
...the positive lead from the charger goes to the other end of the current controller.

You don't need this amount of agitation... just enough to move the solution around.

After 30 minutes the item is dropped into the passivate solution for 30 seconds.

The very first effort came out well and subsequent tests have improved the finish further.

coloured yellow, and it gives the component a tinted finish that, when you see it, will instantly remind you of all the other plated components you've come across before. It finishes the component perfectly and gives it a very OEM appearance.

TROUBLESHOOTING

Incorrect concentrations of the various components in the plating solution obviously will affect how well it works, and increasing or decreasing the electrical current (by moving the crocodile clip along the current controller wire to reduce or increase resistance) should speed up or slow down the process. All companies appear to provide a trouble-shooting guide, with suggestions on how to correct various issues.

RESULTS

As you can see from the pictures, the end result looks pretty impressive , but it's also evident that preparation of the item prior to plating is vitally important. Since this first attempt, Trevor has tried a number of other components, with increasingly convincing results. His tips...

If it goes wrong, you can easily start again after wire brushing or sanding down the part.

As the solution is used, deposits will settle in the bottom of the bucket... if you are using a pump to agitate the solution, this can move the deposits, which in turn contaminate the plated surface of the items. Before you start, filter the plating solution to remove the deposits.

Ideally, get all the parts you want to plate ready at the same time, so that you can do them all on the same day. Setting up and ensuring the solution is working well takes time, so if you can do it all at one sitting, so much the better.

SUMMARY

If the process sounds complicated, in reality it isn't. Some sensible preparation of the various solutions prior to starting should make it a relatively simple job. Expect to test and adjust both the electrical current and plating solution in order to achieve the right results. But once you are happy with the process, the results you achieve can look highly professional.

Rust-proofing

Rust Prevention

Having treated metal with various rust prevention methods, **Ed Morton** now finds out which ones performed the best – and worst!

I was quite gratified by the amount of interest that the rather bizarre looking 'sample stick' received at the recent Stoneleigh show, even though last winter barely happened and so the samples haven't rusted as much as I anticipated. Nevertheless, there are enough differences between the samples to draw some conclusions, and some of the results may surprise you.

To recap, three seperate samples of seventeen popular anti-rust processes, including powdercoating, zinc plating and various paint finishes were prepared, fitted with a rivet and an M5 nut and bolt, subjected to a 'controlled damage' test (using a grubbing mattock…) and then left outside for six months. Obviously, the manufacturers' instructions for each product were followed to the letter.

Fortunately, the results for each sample are remarkably consistent, and, although some of the test pieces showed very little deterioration, the control samples certainly did. This is probably because each piece was prepared with acid etching solution prior to painting. The microscopic surface damage that this process causes gives an ideal surface for paint adhesion, but will also promote rusting. Making a bit of effort with the surface preparation meant that at least some of the products gave very impressive results. There's a lesson there somewhere…

Rusting only occurred at the 'controlled damage' point, so predictably the more delicate surfaces tended to rust more than the more durable ones although, surprisingly, the area around the nuts and bolts or rivets on all the test pieces remained rust free. On some samples the finish started to flake slightly at the damage site to a varying degree, but rusting didn't occur underneath.

The rusting test was scored according to a 1 to 5 scale.
1. No rust on any of the three samples.
2. A trace of rust at one point on the tdamage line.
3. Slightly more obvious rust on less than one half of the damage line.
4. More obvious rust on more than one half of the damage line, but not all the way along it.
5. Significant rust on all three samples along the full length of the damage line.

Damage test was a controlled and consistent drop impact to monitor impact resistence of the various surfaces – and their subsequent ability to inhibit rust.

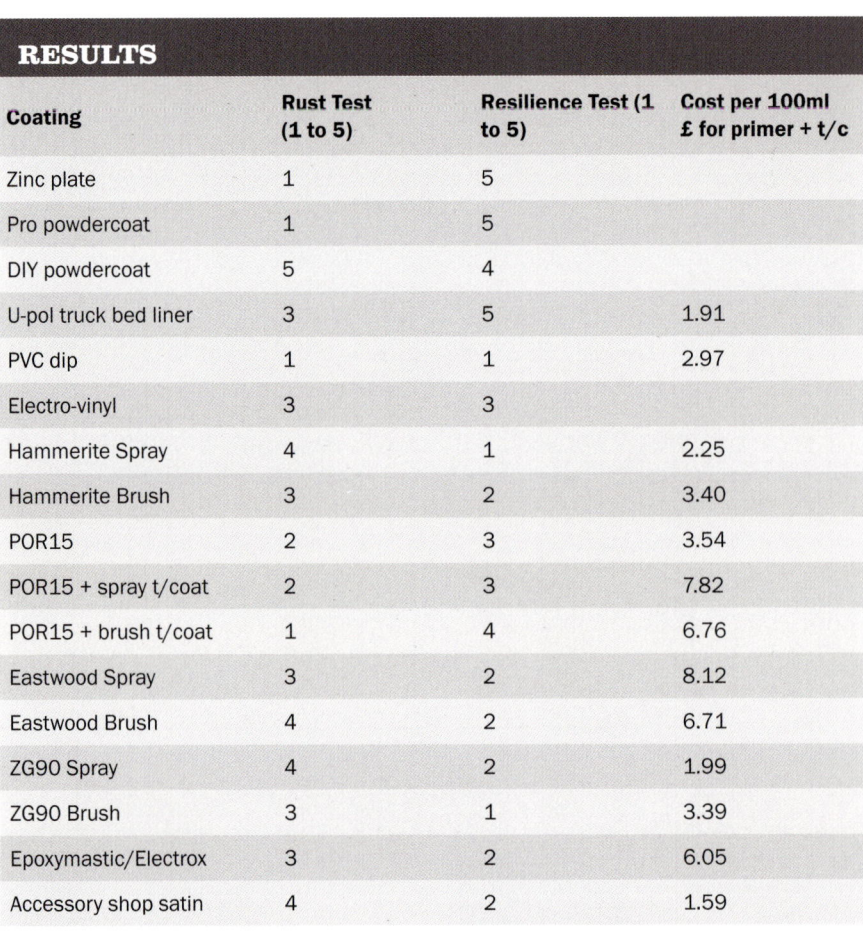

RESULTS

Coating	Rust Test (1 to 5)	Resilience Test (1 to 5)	Cost per 100ml £ for primer + t/c
Zinc plate	1	5	
Pro powdercoat	1	5	
DIY powdercoat	5	4	
U-pol truck bed liner	3	5	1.91
PVC dip	1	1	2.97
Electro-vinyl	3	3	
Hammerite Spray	4	1	2.25
Hammerite Brush	3	2	3.40
POR15	2	3	3.54
POR15 + spray t/coat	2	3	7.82
POR15 + brush t/coat	1	4	6.76
Eastwood Spray	3	2	8.12
Eastwood Brush	4	2	6.71
ZG90 Spray	4	2	1.99
ZG90 Brush	3	1	3.39
Epoxymastic/Electrox	3	2	6.05
Accessory shop satin	4	2	1.59

Bare steel.

Zinc plate.

Professional powdercoat.

DIY powdercoat

U-pol truck bed liner.

PVC dip.

Electro-vinyl

Hammerite spray.

Hammerite brush.

Rust-proofing

POR15.

POR15 plus spray top coat.

POR15 plus brush top coat.

Eastwood spray.

Eastwood brush.

ZG90 spray.

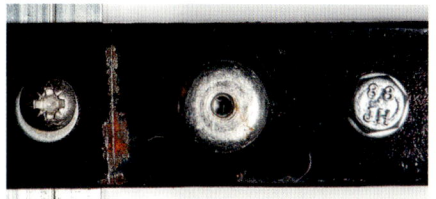
ZG90 brush.

Epoxymastic/electrox.

Accessory shop spray.

The resilience test of the damaged paint edge was judged on a scale of one to five, five being the most resilient, one being the least when attacked with a flat blade screwdriver. This test was more subjective than the rusting test, but in practice the different samples separated into the categories quite clearly when compared to each other.

Samples were scored according to the degree of rusting (see table), and how resilient the paint was around the damage site. Assessment was carried out blindly as far as possible; although some samples are quite distinctive – zinc plating can only be zinc plating, after all.

The professional powdercoating, zinc plating and PVC powdercoating were virtually unchanged after six months outside. The powdercoating remained very durable, but the PVC coating peeled off with minimal encouragement, so isn't really a practical proposition. Damian the metal plater was delighted with the durability of his zinc plating, but he still recommends that it is painted, powdercoated or at least Waxoyled before use. Clear powdercoat or laquer can be used to maintain the technical appearance. U-Pol truck bed liner also gave a very resilient surface, although its performance in the rust test was not exceptional.

The DIY powdercoating was very brittle, and as such rusted more than any of the other samples. However, the broken edge of the coating remained firmly adhered to the metal surface. In an attempt to control the comparison between the types of powdercoating neither was primed prior to application, although this may have compromised the DIY coating more than the professional version.

POR 15 products gave best rust protection of all the paints, and also remained durable for the duration of the test. However, using a topcoat is desirable as POR-15 original is not UV stable. In both the rusting and durability tests the brush-on POR 15 topcoat performed slightly better than the spray version.

Interestingly, and possibly because of the preparation technique, the rust protection and durability performance of standard accessory shop primer's and satin black topcoat was not substantially worse than more specialised and much more expensive products, such as Hammerite, ZG-90, Eastwood Chassis Black and Bilt Hamber Electrox primer with Epoxy Mastic topcoat. The Bilt Hamber Electrox primer's adhesion to the base metal was impressive, but was let down by the subsequent adhesion of the Epoxy Mastic topcoat.

The POR 15 original, ZG-90 and electrostatic vinyl gradually developed a pale surface discolouration, suggesting that these finishes are not stable in UV light, which obviously limits their usefulness.

So can we draw any conclusions from this exercise, or have I simply made three weird metal sticks and wasted some nuts and rivets? It would seem that as long as it's performed well, the industry standard powdercoating for kit car chassis is a good choice, although once damaged, it's hard to repair. Out of the other products tested, POR-15 original with topcoat would be good alternative. Painting a chassis or a pile of suspension components would be a bit more labour intensive, but POR 15 brush paints are very easy and satisfying to use. And although I hate to sound like a pernickety old codger who always measures twice and cuts once, this test definitely underlines the need for good surface preparation before painting.

And, as James Horsley points out, a good smear of Waxoyl doesn't go amiss either.

Thanks to Damian at Vernon Moss Electroplating, Midas Touch Powdercoating in Castleford, and Matt Green for photographing the samples in both of these articles.

Rust Prevention

When **James Horsley** found some corrosion on his buggy's chassis, he knew he had to take action. Time to fire up the Waxoyl pump...

Here's the equipment you'll need for DIY Waxoyling (and the same kit assembled below).

As ever on a kit car, one job soon turns into another. While attending to my buggy's notchy gearchange, inspection of the internal condition of the chassis tunnel revealed some fresh surface corrosion – exacerbated by the return journey from CarFest last year in a full storm I suspect. Having removed various parts and access panels to get to the gear linkage, I was well aware this was an ideal opportunity for some rust treatment/prevention.

If you are running a kit using an older car donor, this process may well be beneficial to you. Vehicles with dedicated chassis (VW Beetle, Citroën 2CV etc) often have key chassis points that can be protected. Modern subframes may also have such maintenance access. If your chassis doesn't have access points, you can drill points at key areas to insert lances, and then fill with suitable bungs afterwards. If welding to original chassis parts you will often not be able to access the rear of the repair for paint, so a treatment like this may be the only option.

There are many solutions on the market designed to seal/treat corrosion, and protect from future corrosion. Some options come with their own applicators whilst others need specialist devices, even compressors. In my stash of parts, I had an old Waxoyl pressure applicator which I originally purchased to protect the inside of my Nova kit car chassis tunnel having replaced the framehead.

This kit comprises a round drum for the Waxoyl (refillable from larger cheaper containers), a pump action head and hose/lance with finer extension. The extension hose is a similar diameter to washer tubing and can be run inside chassis members etc. The end has a fish shaped nail inserted which helps achieve a multi-direction spray when passing through spaces. The main lance also sprays in two directions.

I am fortunate that the beach buggy interior is pretty basic, but if embarking on this on a fully trimmed car I would recommend covering seats/upholstery to avoid any stains from spills. Whilst the applicators will direct flow, I did end up with some Waxoyl escaping from other holes along the tunnel.

MY KIT FOR THE PROCESS COMPRISED:
- A plastic bin full of hot water
- Lots of paper towels
- Applicator pump/nozzles
- Torch

Waxoyl sets to a hard wax which protects surfaces. To apply the grease/wax successfully, it needs to be in a liquid form. I find standing tins in hot water for 10 to 15 minutes with occasional shaking achieves this. Do not try to proceed without softening – your pump will block instantly! As my kit had been used previously, I also soaked the lance and lines in hot water.

I had to transfer Waxoyl to the round applicator tin from a refill 5-litre tin. Once this was done, I stood the applicator drum in my bin of hot water and primed the tin. Pressure is built up by pumping the handle, and then locking down in position. I would recommend

Rust-proofing

Here's the fine lance end for hard-to-reach places!

'Clever nail' sprays Waxoyl evenly.

Fine lance end inserted into the Beetle tunnel.

Tunnel coated with wax internally.

Upper surfaces of the tunnel also coated.

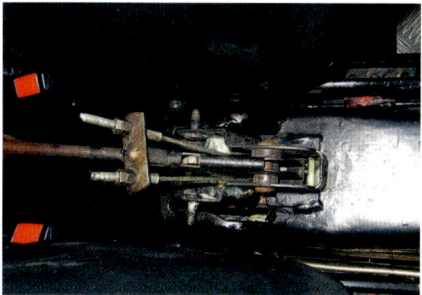
Handbrake was one way of accessing tunnel.

Keep the drum in warm water to make the wax a liquid.

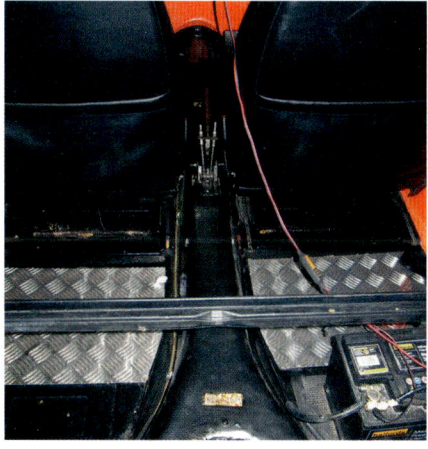
Beetle's central tunnel is the backbone of the chassis.

Access panel re-sealed once an even coat of Waxoyl had been applied inside.

having the lance inserted into your chassis as you prime it. I have found as the lances age that some Waxoyl can escape without the trigger being depressed (my kit is around seven years old).

Once running, the lance releases a fine spray of material in two directions, so with basic movement you can easily coat all aspects of a confined space. I focused initially on those areas I could access with the main lance. When I had covered these, I attached the fine hose, and ran this down the tunnel. Once primed and delivering again, I then slowly drew this down the tunnel. I was able to use the gearshift hole, handbrake mount, seatbelt holes, front framehead panel and gear linkage panel as access points. With around one can (2.5-litre) liberally applied through the tunnel, I hope to have provided some chassis protection to afford the buggy a few more decades use.

Points to note are that this should be performed in a well ventilated area. The smell will continue for some time after treatment as the wax sets. Also, care should be taken with certain painted surfaces. My chassis tunnel was painted with an underseal white spirit based paint, and this was stained by Waxoyl, and loosened paint. Contact with cellulose paints did not cause issues.

Once complete, replace any access panels, or plug access holes appropriately. All parts need cleaning with white spirit or hot water to ensure grease doesn't set in tubes etc. Time invested here will save cursing at the next use.

Current retail prices for the kit vary, but expect to pay around £20 for the applicator, and similar for the grease itself. Shopping around can find bundle kits at good prices. I purchased mine many years ago in a small plastic toolbox, which enables all greasy parts to be stored away together.

Garages will charge anything from £200 upwards for professional application, depending on the amount to be treated and access etc, so if you are prepared to get a bit messy it can be well worth it.

Modify, Improve & Upgrade Your Kit Car

Make A Bracket

You can't build a kit car without making at least one bracket! Technical contributor **Martin Scott** shows you how to add strength into a simple bracket.

Basic tools for the job.

YOU'LL NEED...

- 1.6mm stainless steel flat bar
- Aluminium offcut or similar
- Cold chisel
- Hammer

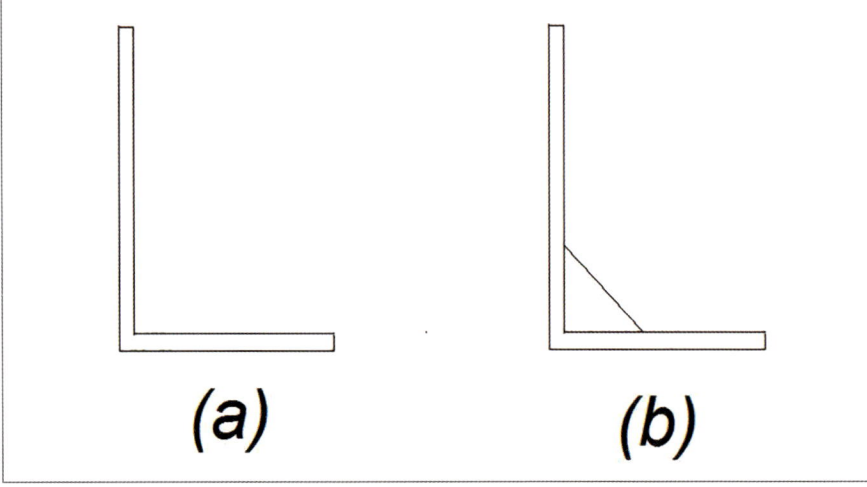

Bracket on the left could weaken and break. The one on the right features additional gusset for more strength.

Fabrication

Flat stainless steel strip is the starting point for our bracket.

Clamped in the vice, additional steel strips protect it from damage from the vice jaws.

Using a strip of steel to create the neat bend in the strip.

Basic bracket now complete. But we want to add strength into the structure.

Fig 1: Using the cold chisel to hammer the bracket into 'V' cutout in aluminium block below.

Fig 2: Here you can see the result of several blows with the chisel.

Fig 3: Flanged edges around bracket add strength.

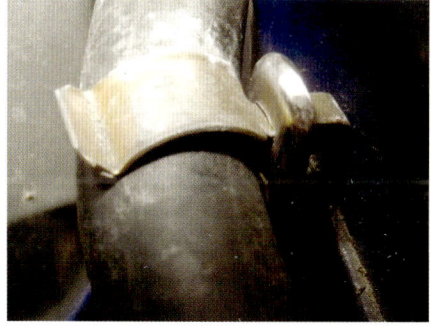

Fig 4: It's designed to hold this Ford hose clip.

Gussetted bracket, ready for holes to be drilled.

We all need small brackets for many purposes on a kit car – mounting wiring clips, attaching trim panels and holding clips for coolant pipes are some examples. Local hardware shops can sometimes provide something to hang a shelf on... but do we want these on our car? What about non-rusting stainless steel? That's more like it.

After a trip to my local steel stockist, I came away with some 1.6mm thickness sheet stainless, and an offcut of aluminium bar. The first job was to make some 'V' cuts in the aluminium with a hacksaw. You probably already have one somewhere in the garage, but with the addition of a small cold chisel, we have our simple tools to make a bracket.

The first diagram (opposite page) shows a simple right-angled bracket, but this could eventually fracture at the bend due to vibration. The second diagram shows the same bracket but with the addition of a bracing gusset across the bend, adding strength. This is what we want to achieve.

Starting with the flat rectangle of stainless, we can hold it in the vice (I've used some steel strip to protect our bracket from damage in the vice), and using another strip of metal and a small hammer, we create our first bend. Figure 1 shows the start of the process to produce the strengthening gusset, and Figure 2 the result after several hammer blows.

The bracket is now ready for drilling (I've marked where the fixing holes will be). And after radiusing the corners and a polish it will be ready to fit.

So why bother with the additional gusset? The main objective is to produce some depth to the bracket, thus keeping it light, but strong.

Figures 3 and 4 show another example where I wanted to secure a Ford metal clip to hold a coolant pipe. A scrap mild steel strip with radiused corners provided the basic former/buck. Stainless sheet was sandwiched between the shape and a piece of softwood. Clamping this hard in the vice pushed the stainless and scrap into the wood, forcing the stainless around the steel former – after some trimming, filing, drilling and polishing it resulted in a strong yet lightweight custom bracket which has now given five years reliable service.

Modify, Improve & Upgrade Your Kit Car

Control Cables

John Dickens shows you how to make up your own cable lengths using the kits and individual components supplied by various companies.

Control cables such as throttle or clutch cables are specific examples of what are generally known as Bowden cables. They are flexible cables used to transmit mechanical force or energy by moving an inner cable inside a tubular outer cable housing. The outer cable is generally of multi-layer construction, consisting of a low friction liner, often Nylon or Teflon, inside a helically wound steel tube with a plastic outer sheath preventing water ingress. The inner cable is usually wound from several thin strands of wire to ensure flexibility.

Traditionally, Bowden cables were used to transmit a pulling force, although push/pull cables have been used as choke cables in the past and have gained popularity in recent years as gear shift cables. Push/pull inner cables need to be more rigid in order to transmit a pushing force without bending and are often made from a few thicker wire strands or even a single wire.

Although fly-by-wire throttle control is gradually becoming the norm for modern production cars, the majority of kit cars still use throttle cables to operate their carburettors or throttle bodies and, due to the alternative engine and driving positions, it is very unlikely that the standard cable from the donor car will fit the new location.

Fortunately, making up a custom throttle cable is not a difficult job and there are quite a few suppliers who can provide kits or individual components to enable you to do the job. Car Builder Solutions has kits for front and rear engined cars and Venhill has universal kits which can be adapted to almost any purpose. Venhill and Car Builder Solutions can also supply individual components should you need a single nipple or length of inner cable.

If you need a one-off cable but are not happy with the DIY option, Venhill will make up one of its Featherlight (yes, I know) cables to a pattern or drawing. These have stainless steel inner cables and are Teflon lined for very smooth operation.

To demonstrate the assembly of a cable, I used a Venhill U01-4-100 Universal Kit which cost £13.37 delivered and is 1.35m long. This kit, which is designed for home assembly, contains a length of unlined outer cable, a length of galvanised steel inner cable and a selection of fittings to suit most applications. In detail, the fittings are a selection of nipples suitable for soldering, a couple of threaded cable adjusters, an assortment of cable ferrules and a pair of rubber sealing boots. To demonstrate the use of solderless nipples I also bought (for £5.44) a cable repair kit from eBay. This contains two gauges of inner cable and a selection of clamp-on nipples.

To save work, the Venhill outer cable already has one ferrule crimped in place and the inner cable already has a nipple soldered in position so all that needs to be done is to cut the components to length, assemble the

The construction of a typical Bowden outer cable.

Bowden inner cable with wound fine wire.

The Venhill U01-4-100 Universal Throttle Cable Kit.

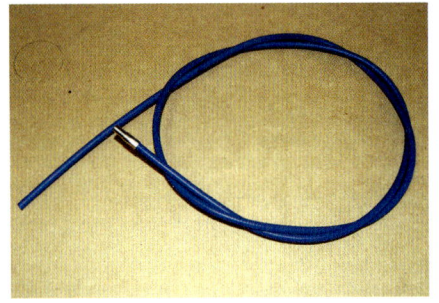
The outer cable is available in a choice of colours.

Inner cable galvanised for rust resistance.

A selection of fittings is supplied with the kit.

Modify, Improve & Upgrade Your Kit Car

Fabrication

These nipples need to be soldered in place.

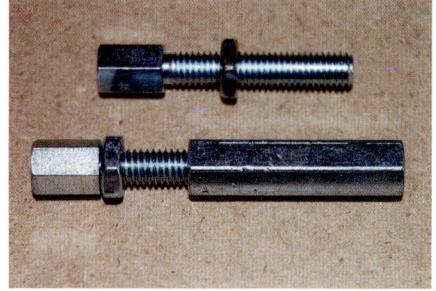
Lower adjuster can be fitted between two cables.

Different end ferrules are supplied.

The rubber boots are normally used on motorcycle applications.

This cable repair kit was a good source of solderless nipples.

These nipples use clamp screws rather than solder to secure them.

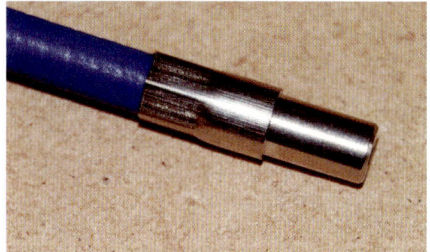
Venhill outer cable already has one ferrule fitted.

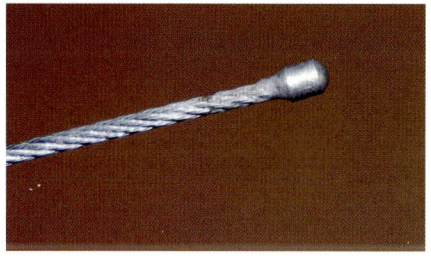
Inner cable has one nipple pre-soldered in place.

Mini-hacksaw is suitable for cutting the outer cable.

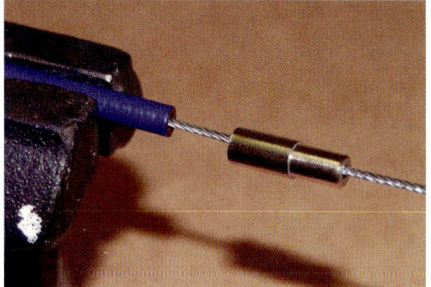
John fits the inner cable before the outer fittings.

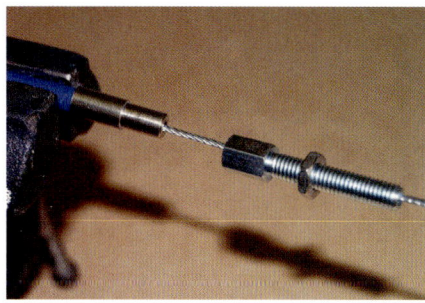
The adjusters need to go on before the nipple.

Solderless nipple can be re-adjusted after fitting.

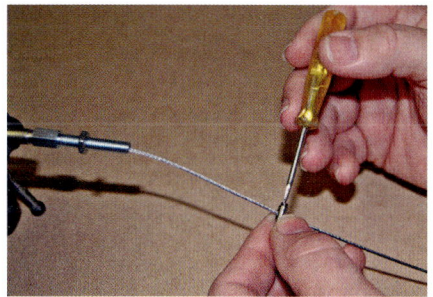
Over tightening screw can damage inner cable.

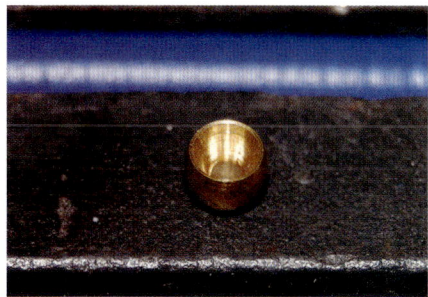
Countersink must face away from the outer cable.

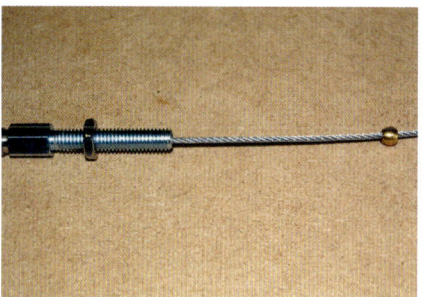
Measure carefully. Once soldered, no adjustment.

Modify, Improve & Upgrade Your Kit Car

Use sharp cutters to cut the inner cable.

Check that all the required fittings are in place.

Spread the cable strands to form a wedge.

Hold the cable vertically when soldering if possible.

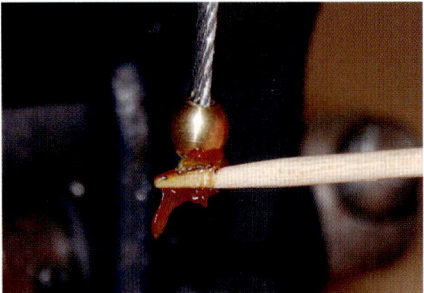
John uses separate flux, even with cored solder.

High power soldering iron will do a better job.

The cavity is completely filled with solder.

Only a short length has been penetrated by solder.

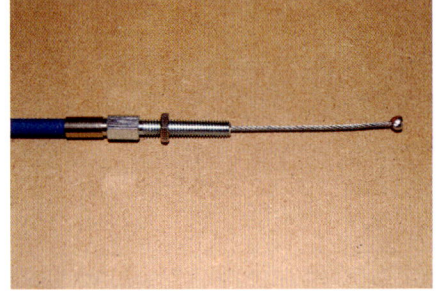
The finished cable is now ready for use.

correct fittings at the other end and solder on the appropriate nipple.

I have found that the best way to cleanly cut an outer Bowden cable is to use a very thin cutting disc in an angle grinder, but failing that I prefer to use a mini-hacksaw. Measure carefully before you cut. Slightly too long is better than slightly too short.

I find it easier to thread the inner cable through next, then to slide on the ferrule and adjusters if required. Correctly crimping on the ferrules requires a special tool but I have found that a couple of drops of superglue or a sleeve of heat shrink tubing will usually secure them in place. They are in compression in use anyway so they will be held in place once fitted. If you are using a solderless nipple, all that remains is to slacken the clamp screw, slide the nipple onto the inner cable and tighten the screw to locate the nipple in the correct place. The excess inner cable can be trimmed off as shown.

Soldered nipples usually have a countersink at one side of the cable drilling. This needs to face away from the outer cable when fitted. Slide the nipple onto the inner cable and carefully measure its exact position. Using sharp side cutters or pliers, cut the inner cable to the correct length. Check the assembly carefully. Beyond this point it will not be possible to fit ferrules, adjusters or nipples if you have forgotten them.

Carefully spread the end of the cable. The aim is to fill this shape with solder to form a wedge in the countersink of the nipple. This ensures that the inner cable cannot pull through. The professionals have a special tool which forms a globe or 'birdcage' shape on the end of the cable to achieve the same effect.

The cable is clamped vertically to minimise the amount of solder which can penetrate into the cable strands by capillary action. This can reduce the flexibility of the cable and cause embrittlement. A small amount of flux is applied to the cable and nipple then heat is applied until the solder melts and runs into the nipple. The solder should tin the frayed cable and fill the countersink to form a solid plug. The solder has penetrated a short length of cable above the nipple but this will not cause a problem. The finished cable is now ready for use. It is a good idea to lubricate it with some light oil before fitting.

In use Bowden cables are generally very reliable but can cease to function smoothly if not lubricated or if water and contaminants get into the housing. If this happens they can be flushed through with brake cleaner or WD-40 and lubricated with a light oil. Modern lined and stainless steel cables are less prone to these problems and Venhill suggests that no lubrication is needed on its Featherlight range although the company still suggests the use of a water displacing aerosol to flush out accumulated debris.

Electrics

Fuel Gauge Fix

Getting a fuel gauge to read accurately in a kit car can be a challenge, but there are lots of ways to avoid the headache of running out of fuel. **Peter Rosenthal** explains.

Most kit cars will use an aluminium fuel tank of an irregular shape. This Gardner Douglas T70 tank uses a VDO dip tube to measure the fuel level, largely because it's baffled with foam.

The earliest generation of VW Beetle didn't have a fuel gauge. Instead owners had to open the bonnet (or boot), unscrew the filler neck and dunk a calibrated wooden stick into the fuel tank. Not very convenient!

Happily, things have moved on now and there are hundreds of different types of gauge and several types of sender to help you check your fuel level at a glance. So if you want to avoid the headache of running out of fuel or having to keep zeroing your trip meter every time you fill-up (and hoping your economy stays constant!) here's what you need to know to get an accurate fuel gauge.

HOW A RESISTANCE FUEL GAUGE WORKS

Most fuel gauges work in a similar way, with some form of floating device inserted in the fuel tank. The floating part is either a sealed plastic float on the end of an arm and floats on top of the fuel (as used in MG Midget tanks as employed by Fisher Furys). The arm is attached to a variable resistor and, as the fuel level drops, the resistance varies. Some senders start at a high resistance with a full tank and then drop to a lower resistance as the fuel level drops. Other senders work at low resistance with a full tank and then increase resistance as the fuel level drops.

The gauge is then matched to work with the sender and displays the fuel level. Where you have problems is when using an aftermarket gauge with a manufacturer's fuel level sender: not all gauges are designed to work with all senders.

OTHER TYPES OF SENDER

VDO has long favoured using a tubular level sender (or dip tube). This encapsulates a floating element in a tube that runs up and down a rod. Two wires on either side of the rod provide the resistance and contacts on the float run up and down the resistive wires to allow the resistance to vary with fuel level. It's claimed to give a more accurate

To check your fuel level sender works, connect it to a multi-meter set to read on the 200ohm scale. The empty reading of this VDO unit is 87.6ohm...

...while the 'full' reading, obtained while tipping the sender upside down, reads 5ohm. This is within the specification of this sender, so it's working fine

Modify, Improve & Upgrade Your Kit Car **69**

reading than a float and arm type sender by being less prone to fuel level fluctuations.

As the moving parts of the sender are inside the tube, these types of sender are the most popular for motorsport use as they allow the fuel tank to be filled with safety foam (which would impede the travel of a float arm sender).

The only point to be aware of is that these senders are less flexible to use as they come in a set length to suit a fuel tank – you can specify them in lengths from 160mm to 700mm. They're easy enough to fit, too – just drill a suitable sized hole in the tank and use the VDO fitting flange (a bezel with five threaded holes and a fuel resistant gasket).

If you want to match it to a VDO fuel level gauge, you'll need to specify either 'dip tube' or 'float arm' at the time of ordering – they're calibrated differently. Typically, a VDO dip tube costs around £50 to £70, with the fitting kit costing £15 and a matching gauge being around £40.

One of the most unusual types of fuel level sender is the ultrasonic type that are usually found on marine applications. They're easy to fit – just drill a hole in the centre of the tank, tap the holes and fit the unit with the gasket – but can be pricey. The one pictured, the BEP TS1 costs £140 from www.tek-tanks.com. The advantage of these units is that they can measure any shape of fuel tank and as they're not immersed in the petrol and have no moving parts, they should prove durable.

Another type of sender that is worth considering is a capacitance type sender, which consists of a rod shrouded by a tube. The sender measures the capacitance between the rod and the tube, which varies depending on fuel level. Commonly used in aircraft, these senders can be cut to length to suit your fuel tank and are said to be highly accurate. They cost around £90 and are easiest to source in America for some reason (ordering via eBay and using PayPal is the safest method of purchase).

Whatever type of sender you opt for, make sure the range of resistance it outputs matches that of the gauge you wish to use.

SENDERS LIVE A TOUGH LIFE – CHECK IT WORKS

Measuring the level of a liquid that's sloshing about vigorously isn't an easy task. Fuel gauge senders spend their entire time immersed in a powerful solvent and do not last forever. So if you're using the donor vehicle's fuel tank, check the sender works before you fit it. The easiest way to do this is to remove the fuel sender with the tank loose on the bench.

Using a multi-meter set to ohms, measure the float gauge at its highest point and at its lowest point. Note down the readings – this will tell you if the resistance is varying in a linear manner and if the sender reads a high or low resistance when the tank is full. Typically, most float fuel gauges operate in the range of 0 to 300 ohms.

If you don't want to remove the sender, attach some terminals to the electrical connector on the sender and attach this to some extended leads on your multi-meter. Note the readings with the tank the correct way up and inverted. It's not as accurate as removing the sender and moving the float manually, but it should give you an idea if the sender is working.

MATCHED GAUGES AND SENDERS

You can't just partner any fuel tank sender with any old aftermarket gauge. Not all fuel gauges work with all sender units and, if you're using a cheaper fuel level gauge that isn't programmable, then you'll need to make sure it's compatible with the sender type you're using. The easiest way to ensure this is to use an aftermarket sender of the same brand as the gauge. This isn't always possible though and, if you're using parts from a donor vehicle and not retaining the original instruments, you'll need to buy an aftermarket gauge.

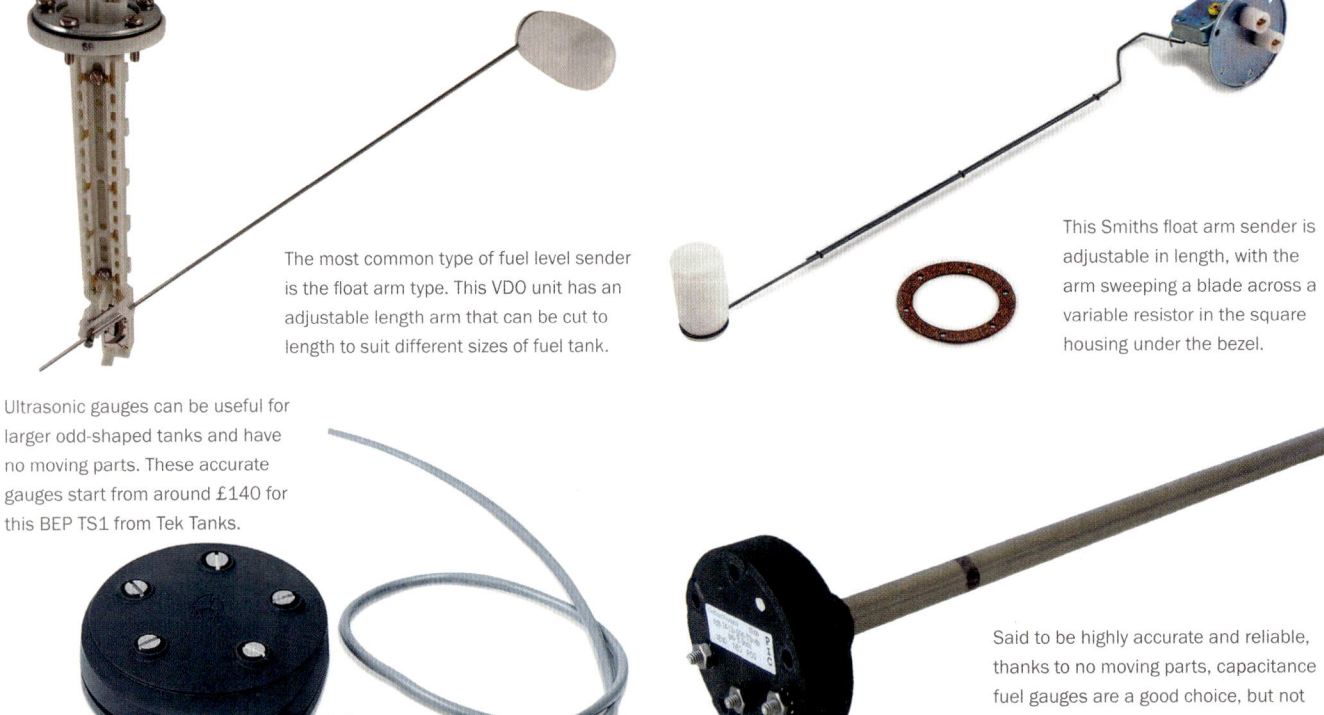

The most common type of fuel level sender is the float arm type. This VDO unit has an adjustable length arm that can be cut to length to suit different sizes of fuel tank.

This Smiths float arm sender is adjustable in length, with the arm sweeping a blade across a variable resistor in the square housing under the bezel.

Ultrasonic gauges can be useful for larger odd-shaped tanks and have no moving parts. These accurate gauges start from around £140 for this BEP TS1 from Tek Tanks.

Said to be highly accurate and reliable, thanks to no moving parts, capacitance fuel gauges are a good choice, but not cheap at upwards of £90.

Top tip here is to avoid going too cheap. There are a lot of unbranded Chinese gauges knocking around eBay at the moment, with prices as low as a fiver. Tempting as this is, having tried these, I'd avoid them like the plague – they're really badly made and tend to come with no instructions, let alone any specifications for calibration. They might cost the same as your lunch, but a sandwich will probably last longer...

Stick to a recognized brand such as Smiths, SPA, Stack, Auto Meter, Racetech, Durite, ETB or TIM and you should be fine.

The easiest method of ensuring an accurate fuel reading is to buy your gauge and sender from your kit car maker. Typically these will be of the same brand. For example, Car Builder Solutions (www.carbuildersolutions.com) offers a Durite gauge with a matching float arm sensor for £50.40 plus carriage.

A recent development from the likes of Koso, Drift and Smiths, and available from Digital Speedos, are self-programming fuel gauges. You fill the tank, press a button, empty the tank and press the button again and the unit callibrates itself. Note, though, that they don't work with VDO senders.

DON'T WANT TO REPLACE THE OE TANK SENDER?

If you are dead set on using your original sender, the first step is to measure the resistance of the sender you have with a multi-meter (see above). Armed with these figures, you next need to find a gauge that is compatible with your full to empty resistance figures. Most decent brands of aftermarket gauge will be able to supply this – go online and Google the fitting instructions for the brand of gauge you want to use and check the specification section, or email the makers.

Non-programmable fuel level gauges, such as this Smiths unit, need to be matched to a fuel sender that outputs the correct range for the fuel gauge.

PROGRAMMABLE FUEL LEVEL GAUGES

The money-no-object solution to matching a fuel gauge to a tank sender is to fit a programmable fuel gauge, such as those offered by Stack and SPA. These are available as individual gauges and are usually built into many all-in-one dash displays. Although the gauges themselves can be more expensive than non-programmable units – Stack 52mm gauges start from £130 – they can save a lot of grief and are the easiest solution of all. More expensive analogue gauges also use stepper motors to move the needle, which can give a smoother movement to the gauge.

The 52mm Stack fuel level gauge works with senders from 0-270ohm, while the 52mm SPA unit works with senders from 50 to 1000ohm. Both gauges are compatible with reversed or normal reading senders.

OTHER USEFUL GADGETS

While you can experiment by adding a variable 0-20,000ohm variable potentiometer (of around 3W power rating) connected between the positive pole of the gauge and the sender input wire, this can be a bit hit and miss. It also won't solve the

One way to solve fuel gauge headaches is to use a programmable gauge such as this Stack version, but they're not cheap – around £130.

issue of a gauge that reads empty when the tank is actually full of fuel, which can happen if the sender and gauge are mismatched. It can also increase the risk of a fire in your wiring loom if you don't know what you're doing so is best avoided.

There are various fuel level gadgets that aim to make your existing gauge work with the donor car's sender.

For example, Car Builder Solutions offer the 'Electronic Fuel Gauge Matcher' for £38.40. This black box is simply wired inline between the sender signal wire and the gauge signal wire input – behind the dash is probably the easiest place to site it. It comes with good fitting instructions and is calibrated for linear tanks out of the box (the instructions also detail what to do if you have an irregular shaped fuel tank).

Another firm offering fuel level correction boxes is Spiyda Design (www.spiyda.com) who offers the Gauge Wizard Mk3 which is also wired inline in the fuel sender wire. It's suitable for senders of a resistance of 0 to 2000ohm (in normal or reversed resistance) and also offers an additional output for a low fuel level light to be wired in. It's priced at £43 plus carriage.

All-in-one dash displays such as this SPA Kit Dash, which costs £839, are the easiest way to sort all your instruments in one hit and can be programmed to suit a wide variety of fuel level senders.

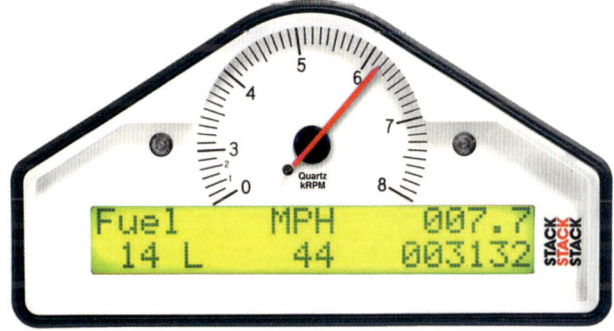

Stack ST8130 dash is another programmable unit and can even display the number of litres left (if correctly calibrated). Costs around £1000.

Sort That Speedo

Getting a kit car speedo to work and be accurate can be a challenge. Here's **Peter Rosenthal's** guide.

Most kit cars tend to use aftermarket instrumentation for ease of build and it can be a challenge to rig up the speedo. If you're lucky enough to be working on a single vehicle donor then you may be able to retain the original gauges, but the vast majority of kit cars tend to use aftermarket gauges for ease and aesthetics. So what types are out there?

MECHANICAL CABLE-DRIVEN SPEEDO
In many older cars, a cable runs directly from the gearbox to the back of the speedo head and directly drives a mechanical speedo. The trouble is that there are lots of different speedo head shapes and lots of different cable ends. If you're lucky and the speedo plugs straight into your donor speedo, you will probably find that the combination of gearbox ratio/differential/tyre size you're using makes the speedo read incorrectly.

Mating up differing types of cable and speedos is easy – simply contact a firm such as Speedy Cables and they'll be able to sort this for you – but calibrating a mechanical speedo can involve either using a different gear wheel in the gearbox drive or getting a specialist to alter the speedo. Either that or you'll need to buy another speedo.

CONVERTING CABLE OUTPUT TO AN ELECTRONIC SENDER
Common kit car gearboxes, such as the Ford Type 9, use a cable-driven output. This can either be blanked off with the cable removed, or you can remove the cable and bolt on an electronic sender unit (some of these stick out so check you have clearance in the transmission tunnel).

For example, Burton Power (www.burtonpower.com) sells a Type 9 gearbox adaptor for around £15 that can be mated up to an electronic speed sensor (£43). Depending on your gearbox maker, some may already have a speedo sensor built into them. Most can then be configured to work with an electronic speedo.

You can also get cable converters that can be added to the speedo end of the cable drive – these are easier to fit than grubbing around under the gearbox. Car Builder Solutions (www.carbuildersolutions.com) sells two types of adaptor that convert cable drives to give a pulsed output: their Digital Speedo Signal Adaptor 1 suits cables of between 2.2mm and 3mm diameter and gives six pulses per revolution at a cost of £22. These can be taped to the end of the speedo cable using self-amalgamating tape.

163mph at 2200rpm... this speedo needs some work!

Electrics

ELECTRONIC SPEEDO

As few people know the precise gear ratios of their donor gearboxes, axles or even what tyre sizes they're going to be using, it's generally a lot easier to opt for an electronic speedo. As these can also be easily recalibrated, they're ideal for future-proofing your car against future upgrades such as changing the gearbox ratios or altering your wheel and tyre sizes.

As there is no physical cable linking them to the gearbox, they're easier to fit too and need less room behind the dashboard.

For all of the above reasons, if you're starting out from scratch it's best to go for an electronic speedo. All versions rely on receiving a pulse signal from an electronic speedo sensor.

Dials and warning lights all in one.

TYPES OF ELECTRONIC SPEED SENSOR

There are two main types of electronic speed senders: magnetic versions and hall effect or proximity sensors. These are generically referred to as VSS (vehicle speed sensors) or as wheel speed sensors.

Magnetic sensors are triggered by a magnet attached to a rotating surface and tend to be small tubular-shaped threaded items that are supplied with circular magnets. These are simply glued on to a rotating surface (bolt head or propshaft body are common places) and the sensor adjusted to a few mm to react to the passing magnet.

Hall effect or proximity sensors don't need a magnet to be glued in place to work, but instead generate their own magnetic field. When a ferrous metal breaks this field, it generates a pulse that the sender transmits to the speedo.

Both types of sensor are available in versions to suit different air gaps (the bigger the gap the more expensive the sensor, typically) and must be matched to the speedo type – some speedos use a magnetic trigger, some a hall effect sender (check the manual or the manufacturer). There are lots of hall effect/proximity senders on the marker (RS Electronics sells lots of different versions) but the key thing is to opt for versions designed to suit the gap you wish to measure and that are unaffected by nearby ferrous metal objects (some senders can't be used if surrounded by ferrous metal). If in doubt, buy the speedo maker's sender.

If money is no object, the easiest thing to do is to buy a new aftermarket speedo and a matching sender. All you need then do is fabricate an attachment bracket (strip aluminium from B&Q is ideal for this and doesn't rust or fatigue).

FRONT OR BACK WHEEL, OR PROPSHAFT?

Most car makers build the speedo sensor into the gearbox, but if you're not using the original cable or electronic sender, then you can choose to use front or rear wheels or even the propshaft or CV joint bolt heads.

The best advice is to use the easiest place that will be affected by heat, dirt and debris the least. For example, Westfield fits the speedo sender on a bracket pointing at the propshaft bolt heads on its independent suspension rear cars, while on something like a Fisher Fury it can be easier to use a sender on the front hubs, aimed at the rear brake disc bolts.

Bear in mind that brake discs get very hot and can expand if you use the front wheels and factor in over-reading due to wheelspin if you use the rear wheels.

GPS SPEEDOS

All manner of GPS-based speedos are available, ranging from free apps for your smartphone (or satnav) to standalone units that can be attached to the windscreen on suction pads and plugged into your 12V accessory socket. Certain brands of electronic speedo also have plug-in GPS senders that can be discretely positioned on the dash-top.

All-in-one Stack ST9918 is a dream system – but it's £3100!

Modify, Improve & Upgrade Your Kit Car **73**

Many modern 'boxes such as this Tremec have an electronic speed sensor that can be calibrated for an electronic speedo.

For Type 9 'boxes, adaptors like this allow an electronic speed sensor to be attached to provide a pulsed signal.

To save working at the oily end of the speedo cable, CBS offers an adaptor to convert the cable's rotation to a pulsed output.

This RS Electronics proximity sensor detects ferrous metal parts passing its edge – it gives a pulsed signal on detection and also lights up an LED to indicate wheel rotation.

Digital Speedos sell these magnetic speedo sensors as part of a package with their speedos – they simply plug in to the back of their units. Simple!

This magnetic speedo sender has been mounted to a Sylva Stylus on a homemade aluminium bracket and detects a magnet that has been glued to the head of a brake disc bolt.

These are probably the easiest types of speedo to set up, but before you buy one be aware that none will pass the IVA test. It's very clear: "The types manufactured for bicycles, racing only, those that rely on GPS or those that require switching as a separate function to that of operating the vehicle or those that have a separate power source from the vehicle or where they do not operate as an automatic function when the vehicle is driven are not permitted."

Equally, a keen-eyed MoT tester should also fail a car that uses a GPS-only speedo that doesn't display a reading when the vehicle is indoors (as it can't see the satellite). They can also give no reading in tunnels and certain brands have been criticized for having a poor refresh rate. Given how light and rapid most kit cars are, a laggy speedo isn't ideal!

CALIBRATION
Apart from GPS speedos and cable speedos, most electronic speedos need to be calibrated before they can read accurately. This either involves setting the unit in a programme mode and driving a set distance (eg a mile – measure it in your daily driver first and mark the end point by a landmark or with a rock) or by entering some calculated values such as the outer circumference of the tyre. While tyre makers do quote the outer circumference of the tyre online, it's best to measure it yourself by simply wrapping a tape measure around the outside of the tyre and entering that figure.

Once the speedo is set, check its accuracy with either your satnav or a smartphone app. Make sure it doesn't under-read. If it does, you need to reprogramme it – legally speedos can over-read by a small percentage but must never under-read. Bear in mind that as your tyres wear down, the speedo will naturally over-read as its rolling circumference reduces. For a given mile, a smaller wheel will rotate more, increasing the speedo reading.

VERDICT
If you're starting from scratch, the easiest thing to do is to buy a speedo and sender as a combined package – that way you can be sure they're compatible and will work seamlessly. Watch out for earthing points during fitting – many need to use a common earth point for speedo and sensor and keep the speed sensor wires away from any sources of electrical interference for maximum stability. Once you've fitted the speedo, check the calibration with satnav or GPS speedo on your smartphone and adjust it accordingly.

Electrics

Headlight Upgrade

Are your headlamps like glow worms in a jar? Then you might need to upgrade your headlamp wiring. **Peter Rosenthal** shows you how to do it in one afternoon.

If the headlamps on your kit car are disappointingly dull, there are a number of things you can do to get brighter beams without breaking the bank.

It's all about the battery. Before you invest any money, though, get your multi-meter out and check the voltage at each headlamp connector. A fully charged battery is typically around 12.7v to 12.8v (or up to around 13.5v if the engine is on and the alternator is working). 12.5v means the battery is half flat and 12v means it's flat. So, assuming your battery is fully charged, you should see around 12.7v at the headlamp connector. If you don't, there may be room for improvement.

Sealed beam or halogen? If you have an elderly kit car that's still using sealed beam units (where the entire headlamp is the lamp), the first step is to remove them and lob them in the nearest skip. Various firms sell halogen upgrade kits and if you Google you'll soon find a set. The 7in set featured here cost £28.60 from ebay supplier Mk1ClassicCarParts (the eBay moniker for Vintage Warehouse 65 Ltd).

Are your lenses milky? Just like a pensioner suffering from cataracts, you'll never get a decent light output if your headlamps aren't up to scratch. If your reflector backing is corroded on a glass headlamp, there's no option but to replace it. Equally, if you have plastic headlamp lenses that have turned a bit milky then it is sometimes possible to polish them. There are various plastic headlamp restoration kits on the market and I've had good results with the Meguiars One-Step Headlamp Restoration Plus kit (which is actually about three-steps!). Some people also advocate using fine wet and dry paper or even a mild abrasive like toothpaste, but be wary as you may remove the UV coating of your headlamps. However, if your plastic headlamps are milky, you haven't exactly got much to lose...

HID upgrade kit. China has flooded the market with cheap High Energy Discharge headlamp upgrades and eBay is awash with them. The trouble is the quality can vary enormously and, more importantly, they're not legal for road use and do not have a cat in hell's chance of passing an IVA test.

While some versions may pass an MoT and may do no harm, the fact remains that if you have any accident at night you risk the wrath of your insurance assessor. If in doubt, they simply won't pay out (and don't kid yourself that insurance assessors won't spot them, either – they're trained specialists).

There have been many reports of premature failure of these units and poor beam patterns. As Xenon headlamp kits typically run at 65v, there is also a much greater risk of electric shock and fire risk.

The low price of these kits is tempting, but none are E-marked for use in a halogen headlamp unit and none are road legal in the UK. Avoid.

LED headlamps. While LED tail-lights have been around for years, LED headlamps have taken far longer to get to market and are only just starting to appear on the aftermarket.

While you can get a number of E-marked units, they're massively expensive – typically around £400 a pair.

Replacement LED bulbs are also available in common sizes (typically H4 for 7in halogen headlamps) but these items are not E-marked and therefore not road legal or capable of passing IVA. They may also cause 'bulb failure' warning lights to illuminate in more sophisticated wiring looms.

One of the issues with developing replacement LED lamps has been dissipating the heat these bright lamps generate and many use a large aluminium heatsink on the rear – this may cause fitting issues if your car uses them.

As with aftermarket Xenon HID units, using aftermarket LED bulbs risks invalidating your insurance.

IVA and MoT. IVA requirements state that any lights must be E-marked and that headlamp washing systems are not needed for either HID or LED lamps provided the light output is below 2000 Lumens (you have to prove this with documentation). They also have to pass strict beam-pattern tests.

Modify, Improve & Upgrade Your Kit Car **75**

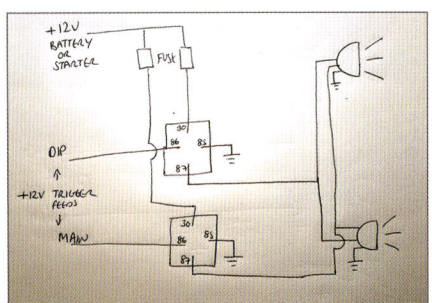

In this case, two fused relays were used for simplicity, with one controlling the dipped beam and the other relay the main beam. Both take power from main terminal on the starter motor for convenience (battery is in the boot). The 12v feeds for the dipped and main beam are taken from the original headlamp feed wiring, but you could take them from a switch or the stalk.

This is the kit you need to do this task – some 27amp cable for the main headlamp feeds (I used about 5m), a couple of relays, wiring, ring and spade terminals, conduit and, if needed, some new halogen headlamp bowls. Tools needed include a multi-meter, a selection of screwdrivers, soldering kit and some wire strippers and crimpers.

Remove the old headlamp bowl. Throw it away if it's a sealed beam! If it's not, Unscrew the headlamp bowls to access the wiring to the rear of the headlamp bowl (as in the main image above).

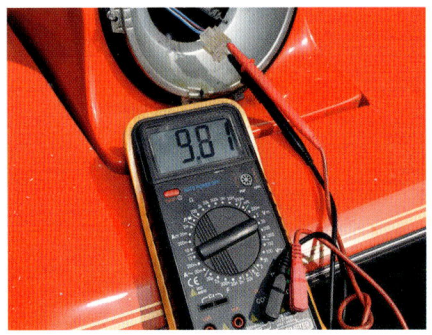

Before you do anything else, measure the voltage at the headlight plug. If it's a good healthy 12.7v or more, you're not going to get much better so stop reading now! In this case the lights were only outputting a feeble 9.81v. Much of this is due to the path the wiring takes and perhaps the age of it in this case. Clearly room for improvement!

Use the multi-meter to identify the dipped beam feed current (blue with red stripe in this case) and the main beam feed current (blue with white stripe). Label them if you're forgetful...

Under the bonnet, measure the voltage either directly at the battery or at the starter motor. This Lotus has the battery in the boot, so the starter motor was far easier to get at. Find a good earth (I used the engine block) and measure the voltage. This is what we need to see at the back of the headlamp plug for maximum efficiency.

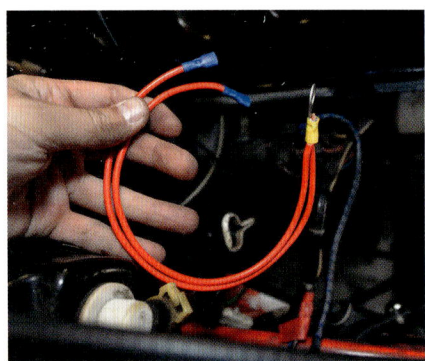

Using two sections of 25amp cable (or something similarly meaty) crimp on a ring terminal and cut the cable to the position of your two relays. I used fused relays mounted as close as possible to the starter motor wiring. If you don't use fused relays you must fit an inline fuse to each wire or run each through a fuse box. Remove all the fuses until the wiring is complete.

Identify a convenient earth point. Check it with the multi-meter. This is very important and will save you all the usual 'why isn't it working?' headaches down the line that are invariably caused by dodgy earths. Make up two earth wires with ring terminals to this point and feed them through to your relay point.

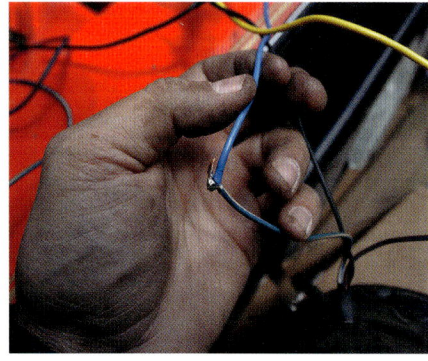

Feed a new 27amp wire (or similar) to the back of the headlamp – this is the new main beam feed for the headlamp connector. Solder it in place and wrap connection with insulation tape. Ideally you'd use headlamp connectors with a thicker wire but the original items were in good condition and the voltage drop in the few inches of thinner gauge wire shouldn't be too great.

76 *Modify, Improve & Upgrade Your Kit Car*

Electrics

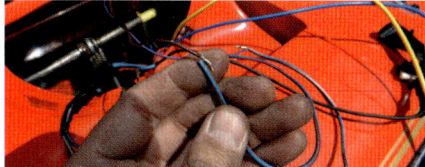

This brown and yellow wire is attached to the original headlamp dipped feed (you could also use the back of the stalk or another convenient point in your wiring loom) using solder and then taped up. This wire is then fed through to the dipped beam relay under the bonnet to energise the relay. It's not carrying the main lighting load so doesn't need to be of a heavy gauge.

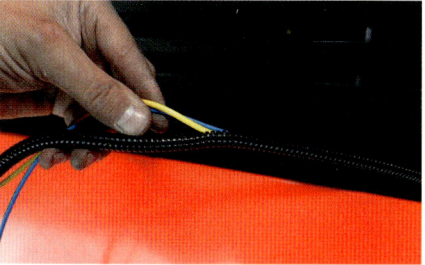

A yellow dipped beam wire and a blue main beam wire were then fed through to the other headlamp connector and, after cutting and sealing off the original wires, were soldered to the back of the headlamp connector wiring.

All the power feed wires were fed back to the two relays and attached to them with crimp-on female spade terminals. The red wires from the starter motor connect to position 30, the earth is position 85, while position 87 is the new high-power feed wire to each of the headlamps. Position 86 triggers main or dipped beam on each respective relay.

Attach the relays in place – in this case existing screws were used to avoid drilling extra holes (I may use the spare positions on the fusebox and get some unfused relay later). Place the fuse in place and check your wiring.

Now to see if all your hard work has paid off. In this case I was delighted to record 12.80v on dipped beam on the offside headlight...

...and 12.87v on the nearside headlight. On most kit cars you probably won't see such a dramatic increase in voltage, but you will see an increase as there is less resistance in the circuit and a more direct route from the 12v power source.

As no HID kit or replacement LED bulb is E-marked, both will fail the IVA.

However, complete LED lamps that pass IVA are available – but you must make sure that they are E-marked, they work in the precise beam patterns tested at IVA and the manufacturer supplies paperwork to document that they are under 2000 Lumens.

For the MoT, the manual simply states that if HID or LED lights have a headlamp washer system or a self-levelling system fitted then it has to work. Equally the lights have to pass the beam pattern tests.

Upgraded halogen bulbs. There are lots of aftermarket lamps claiming all sorts of wonderful benefits and with prices ranging from £15 to £60 a pair. Having tested a £60 PIAA set a few years ago (and returning them as they barely made any difference) the best bet is to stick to a good pair of branded bulbs from GE, Bosch, Philips, Ring, Osram or Halfords. A good tip is to see which lamps Auto Express recommends – they test them annually. Their current winner is the Philips X-treme Vision (£16 a pair plus postage on Amazon) followed by the Osram Night Breaker Unlimited (cheapest price £11 a pair plus postage on Amazon).

Upgrade that wiring. Typically, kit car headlights are fed by tiny wires that look more like parcel string than anything else. If you've upgraded your halogen lamps and want more, then a simple and effective upgrade is to trigger your headlamps via a switched relay that uses larger capacity wires to feed more current to your existing lamps for greater brightness. It's a trick classic car owners have been using for years and it works brilliantly (sorry!) on kit cars. It's IVA-friendly and road legal.

The wiring is similar to that for spotlamps and uses the original dipped beam feed to energise a relay which then feeds the headlamps via a large diameter wire from either the battery or the starter motor feed – whichever is handier. A second fused relay is used to power the main beam.

If you've built a kit car you'll probably have all the bits you need to carry out this upgrade gathering dust somewhere in your workshop and if not it'll only cost you around £20 to £30 to get all the bits needed from your local motor factor or a kit car show.

In this case I carried out the headlamp conversion to my dim-headlighted Lotus Elite. OK, it's not a kit car but it is made of fibreglass and isn't far off one!

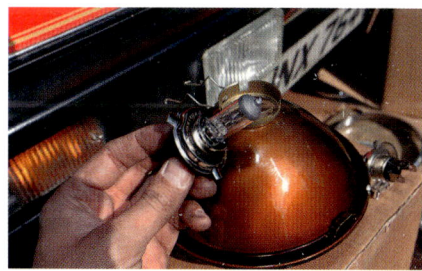

With power to the connectors checked on both dipped and main beam all that's left to do is fit the new halogen bulbs into the new headlamps and reattach them. Don't touch the glass part of the lamp (but you knew that, didn't you!).

With both headlamps in place, it's time to test them. Brilliant! The chromed bezels have been left off to give access to the adjustment screws – it's a good idea to visit a local MoT centre and have them set up the beam pattern for you.

Modify, Improve & Upgrade Your Kit Car **77**

Donor Car Looms

Martin Scott looks at the basics of removing a loom from a donor and refitting into your kit car.

Now unless you've taken up the latest teenage craze of making things with small elastic bands and have no interest in kit cars, then your 'loom' work will focus on wiring... Fearful to some, and a mystery to others, but not too bad if tackled in stages!

How many kit cars look great until the bonnet is lifted and a tangle of wires is revealed? I remember my first kit car (a Dutton Melos) where to 'lose' some length in the wiring, I curled it under the dash in a loop! Thankfully, I progressed from that poor start. So for this article we'll be looking at removal from a donor car, renovation and preparation for placement into your kit car.

REMOVING THE LOOM

Before taking any parts off your donor, label any wiring connectors carefully with masking tape and pen and take photos as well if necessary. It's all too easy to think "I'll remember that" to find several weeks (months?) later that it all looks less memorable.

Most plastic connectors in a modern car will only fit one way, and to one item (the process is called poka-yoke), but if in any doubt, label it well. It would be possible, for instance, to get the lighting part of the loom transposed, with the wiring for the left-hand indicator attached to the right – a bit embarrassing to say the least!

If the connectors have plastic retaining clips holding them to the components, be

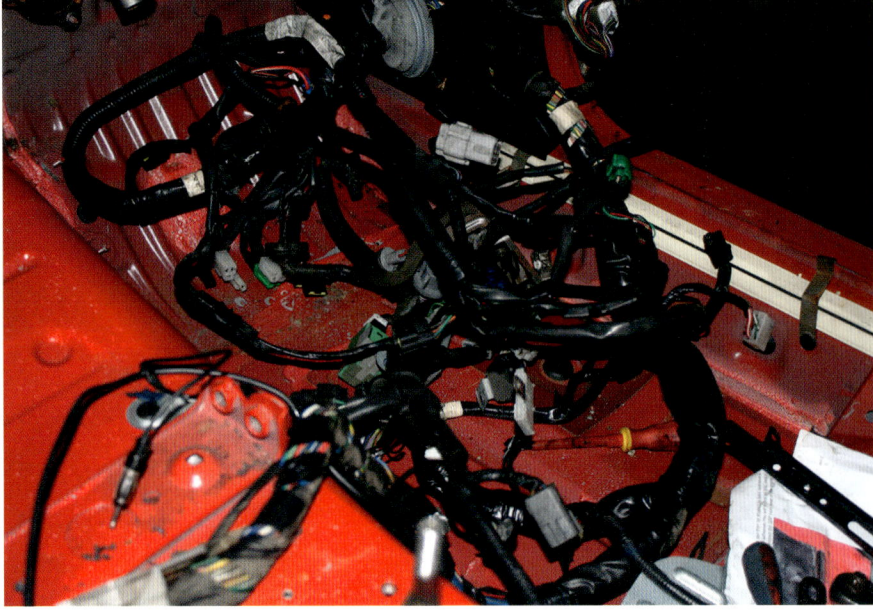
Donor looms can be large. Label everything, as you won't remember later.

gentle when removing them so the clips don't get broken – under-bonnet heat can make them brittle. With all the connectors removed from the components, it is then a case of removing the loom from the donor car. It will probably be held in place by clips which peg into holes in the bodywork, cable ties with attaching legs or other similar means. There may also be earthing tags or eyes held with screws to the steel. Whatever method is used, it's best to keep all the fixings in case you need them for your kit car.

RENOVATE

Now the loom is removed, sit down on your best sofa (OK, maybe not indoors unless your other half is very understanding) and carefully clean the loom and connectors with a damp cloth. Examine every plastic connector housing and the metal connections (terminals) inside. You are looking for any damage/cracked plastic, and any corrosion on the metal terminals. If the corrosion is minor, then removing it with a fine wire brush

Connectors often have plastic retaining tags... take care when undoing them, as heat from the engine bay can, over time, make them brittle.

Retain the loom retaining clips from the donor car, as you may be able to re-used them...

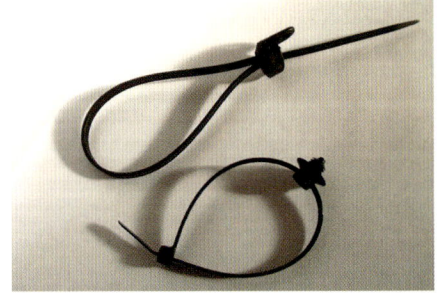
...although the cable ties cannot be retained, the mounting clips can.

78 Modify, Improve & Upgrade Your Kit Car

Electrics

It's worth holding on to any loom earthing eyes.

Check connectors and terminals for damage...

...and replace any that look suspect.

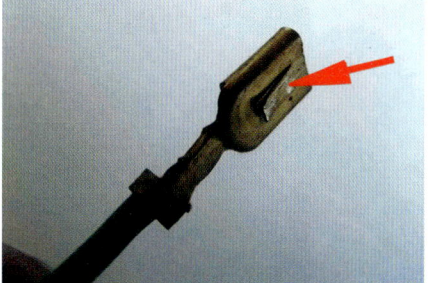
Metal terminals typically held in with metal tags...

...With a little imagination, you can make up release pins such as these...

...Gently push them under the terminal while pulling the cable from the connector block.

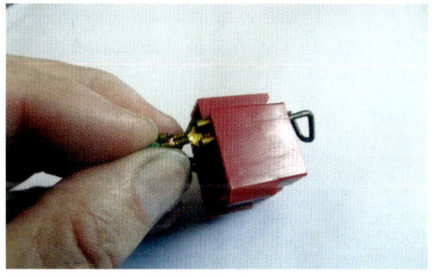
...You'll find the clips will then pull out quite easily and can be re-used.

Some multiplugs have numbers to help ensure cables go in the correct position. Make a note of the numbers before removing the cables though!

Holding the connector this way ensures the solder only flows slightly into the wires.

may suffice, but if it has badly corroded then it will need replacing.

Getting the metal terminal out of the plastic connector housing can involve a bit of investigation. Most modern ones are held in place by a plastic latch on the housing springing into place, or part of the metal terminal springing into place in a similar manner. Careful examination will reveal which part needs to be manipulated, and this can be done using a sharpened piece of spring wire or ground junior hacksaw blade (see Fig 9 for some I made). Just push the tool into the connector whilst pulling the wire gently and hey-presto!

Replacement connectors can be sourced from various kit or classic car electrical part suppliers. If you are unable to find a replacement, then now that the connector is removed from the plastic housing, it can be carefully cleaned, heated with a soldering iron and tinned with solder to minimise any further corrosion.

If replacing with new, then cut back 25mm, bare the wire of insulation and fit the new terminal in place (not forgetting any waterproof seals which will need to be put on the wire before the terminal!). If each terminal is done individually and replaced before another is removed then the position in the housing will be correct. If removing them all from the housing before working on them, either draw a diagram showing their locations/colours, or label the housing with masking tape noting the wire colours. As an added bonus, the plastic housing is often marked with numbers for each terminal.

The terminal can be attached to the wire by crimping, or by crimping and soldering (my preferred method). If soldering, heat the terminal with the soldering iron, and apply resin-cored solder. This is best carried out with the wire vertical and the terminal at the lowest point – gravity assists by minimising solder flow into the wire, which would create a stiff portion of wire which may break with vibration.

LONG OR SHORT

Shortening a loom is easier than lengthening it. The best way of going about this is to offer it up to the new chassis and decide what outcome is desired. For example, if you have a hinged rear clam, and some length has to be removed from the rear loom, then this can include new multi-way connectors giving the added benefit that if you ever need to remove the clam, the wiring can be easily disconnected and reconnected.

Modify, Improve & Upgrade Your Kit Car **79**

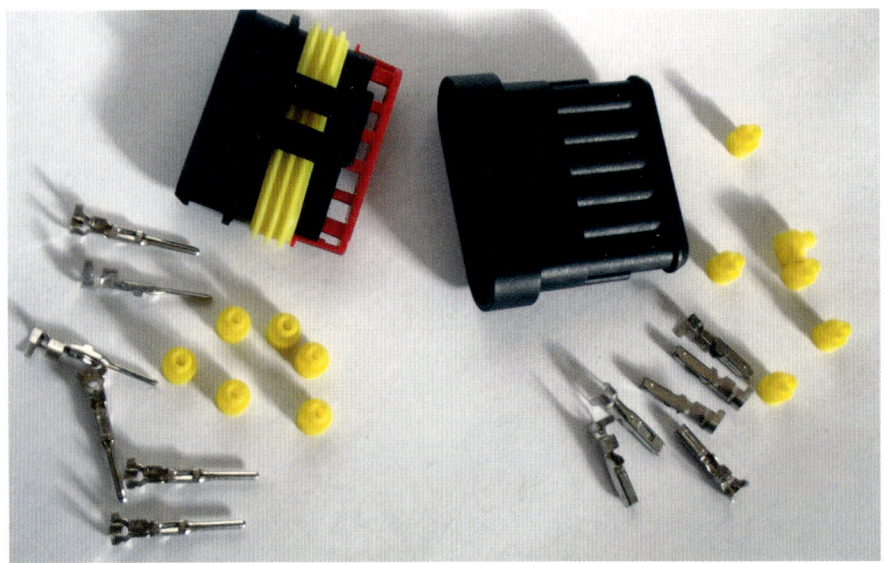

Multi-way connectors such as this are ideal for disconnecting a loom from a removable panel.

If the loom has to be shortened and is unlikely to need removal, then cut out the required length, slip some heat shrink sleeve over one of the ends, solder the ends together in line, put the sleeving in position, heat and the job is done! A neat tip is to stagger the joins on neighbouring wires to avoid a large 'lump' developing in the loom.

Lengthening the loom presents the problem that extra wire is needed (of the same colours and rating), and also many joins mean a lot more work. One way around this is to get another loom from the same model of donor – this is especially useful as it reduces the number of joins. Got a kit car with a long bonnet? No problem. Just use the front loom of a second car and join it to the first, and all the wires are correct and the length can be almost twice as long!

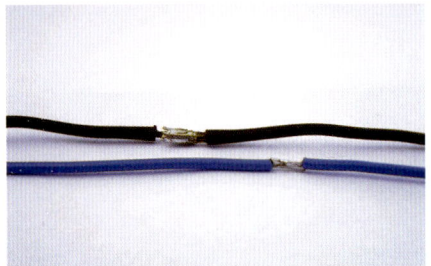

Staggering joins in multiple cables reduces the diameter of the loom in this area...

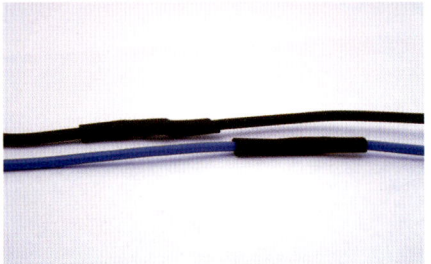

...Heat shrink sleeving keeps everything tidy and safe.

PVC tape is available everywhere, but is not ideal for covering large lengths of cable...

...much better is non-adhesive loom tape or, as seen here, self-amalgamating tape...

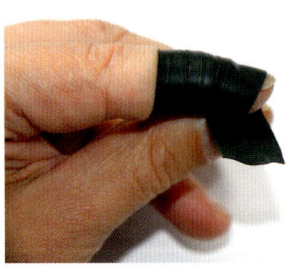

...Self-amalgamating tape fuses together when wrapped around on itself.

Split convoluted tubing is ideal for covering a donor loom where multi-plugs etc are already in place.

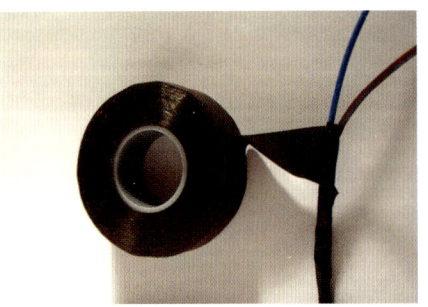

PVC tape can be used at the end of a run of loom tape, to stop it unravelling...

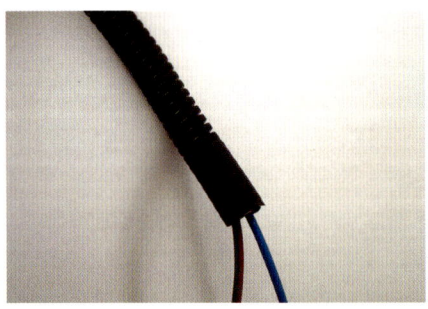

Split convoluted tubing certainly gives a very tidy finish.

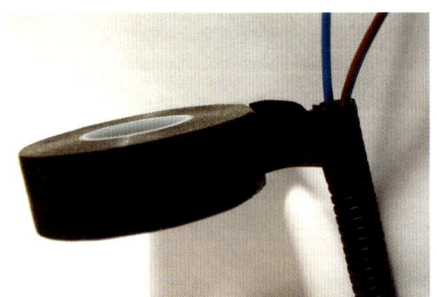

...it can also be used to 'close' split tubing ends...

These natty unions allow for the introduction of T-pieces into existing tubing...

...These Y-pieces achieve the same function and give a professional finish.

Clips from donor, loom covered in tube and finshed with a plastic end piece, into donor multi-plug.

NO STEEL CHASSIS

Most kits have a steel chassis, and any earthing eyes can be attached to the chassis by drilling and tapping into a suitable place just like on the donor. If you have a GRP monocoque kit such as a Midas, GTM or Quantum, then you will need to add earth wires to the loom. This has to be done with care to ensure it's up to the task, and is best left to a separate article.

INSTALLATION

As far as I know, there aren't multi-coloured loom coverings available! Any colour as long as it's black seems to be the way and there are a number of options:

Firstly it can be left as original, and if that is aesthetically acceptable then no problem. However, if it's been lengthened it will need covering in some way. PVC tape should be avoided (although it has those pretty colours) because it can get horribly sticky. A better option is non-adhesive loom tape, which gives a nice uniform finish and can be simply wrapped around the loom. Better still, is self-amalgamating tape, which although not tacky, fuses with itself to give a watertight seal, and doesn't tend to unravel when the loom is moved around to put it in the car. Most of us will be happy with self-amalgamating where the wiring isn't visible.

Which leads us onto wiring that will remain visible (such as in the engine bay). PVC sleeving or convoluted tubing which doesn't have a split gives the problem that it has to be put on before the connector housings, or all the terminals have to be removed from the housing so you can (try to) slip the wires through the sleeving – a bit bothersome when the donor loom has all the connectors/housings in place!

Luckily help is at hand in the form of spiral binding, or my personal favourite... split convoluted tubing. Both can be put in place without removing the connectors.

The split convoluted tubing gives a great, easy-wipe clean, finish, and can be additionally secured with a little PVC tape which can be wrapped around the loom, through the split and around the sleeving. For a little extra cost, a really professional look can be achieved with the addition of clip-on ends, Y-pieces and T-pieces.

Be careful with wiring which may get hot or contaminated with oil vapour – either use the donor car's covering in susceptible areas, or play safe with a high temperature/contaminant covering such as the expandable glassfibre or polyester sleeve, which can be stretched a little to allow the connector housings to pass through.

SUMMARY

Plan this part of your project well, and the results can be very satisfying, combined with the added reassurances of reliability and safety.

Take a methodical approach to labelling, making notes and deciding where to introduce joins and multi-way connectors. Together with replacing or renovating any corroded or damaged parts as you go along, the results should be impressive.

Don't forget the IVA requirement of securing the loom at intervals of less than 300mm, and ensure it is unable to contact any moving parts.

As they used to say on Blue Peter, here's one I did earlier, compared with a rather ragged alternative.

Original bits of donor sheething, loose cables and a distinctly messy end result. You can do better!

Floor A Garage

Want to upgrade your garage? **James Horsley** outlines your re-flooring options – before choosing and installing his own.

Painted finish cuts down on the dust of a concrete floor. Easily damaged, though.

When kitting out a garage or workshop, often the focus is on lighting, tool storage, power supplies, maybe even running water. Flooring is often overlooked, but can make a real difference when working on your car. Any basic garage is likely to have a bare concrete floor. Such floors can be cold, damp and inherently dusty. As you're likely to spend a lot of time on the floor, this article is intended to outline the options to you to ease your build (and back!). In reviewing the options available I have also been considering my own workshop floor, which is in need of attention.

CONCRETE FINISHING

A bare concrete floor if unsealed will generate dust which can be abrasive to paintwork and gelcoat, and also prove annoying when sticking to bolts and other greasy parts you may drop or place on the floor. A simple basic step that can help is to seal the floor with a suitable concrete sealer. These are available from most DIY stores and are simply brush or roller applied. Instructions should be closely followed for new concrete finishes, as these should not be painted or sealed until fully cured.

If your concrete floor is damaged you may wish to repair areas prior to sealing. Mild damage/irregularities can be repaired with screed mixtures which are a combination of a latex adhesive and cement. These are typically self levelling, but only work in shallow repairs. Screed compounds can be bought from DIY stores as a dry powder, typically in 10 to 20kg bags for mixing at home, though smaller (but often more expensive) patching kits can be purchased.

Deeper repairs will need a cement patch. My own floor is concrete with an epoxy coating, that was fitted when new in 1968. Various areas of this coating have deteriorated and I have had to patch these with a screed repair. The best repairs were achieved when I had removed more of the damaged area – it is important to ensure any loose material is removed for bonding.

Expect to pay: £20 to £30 for a 2.5-litre tin. Coverage will vary. One to two tins will cover a typical single garage.
Pros: Simple to paint, low cost.
Cons: Need to clear floor, doesn't change finish though does typically darken concrete slightly.

Concrete floors prone to damage. A screed will work as a fix before you begin reflooring.

Repaired floor ready for tiling or painting. Pros and cons of both options outlined in text.

In the garage

A typical painted garage floor.

Painted surface helps cut down on the moisture inherent in a concrete floor.

Altro flooring can be 'welded' together where it joins for a professional look.

CONCRETE PAINTING

Sealer coats can also be used in preparation for painting concrete floors. Concrete paints will be absorbed more in an unsealed floor, increasing paint consumption and cost. Floor paints are available at most DIY stores and motor accessory shops, though often only in 2.5 to 5-litre tins. Buying online can be more cost effective with specialist suppliers offering paints in sizes up to 25 litres.

Floor paints can be purchased in a wide variety of colours and transform a garage greatly. However, stating the obvious, painting can be disruptive as the area needs clearing completely to paint effectively. If you chose to push all of your stored items to one end and paint in sections it is hard to achieve a smooth finish that doesn't look patchy, and it does of course lengthen the time taken. Temperature is also a key factor; curing can be delayed greatly in cool environments.

Paints often have rubber or epoxy additives in them to assist with resistance to damage. If regularly parking a car in the same area, damage can occur at pressure points – tyre patterns are often visible after prolonged use. Certain chemicals, such as petrol and brake fluid, can also damage paints so care is needed when undertaking certain tasks. If using heavy lifting equipment, metal wheeled trolley jacks etc damage can also be caused, protective mats can be useful to avoid this.

I have had reasonable success with a painted floor in my garage over the last seven years, having repainted twice in that time. I also chose to paint up the wall by around 20cm to help with mopping/washing the floor down. I have tried to use offcuts of vinyl or carpet under trolley jacks to reduce damage. Over time, though, laziness can slip in! Brake fluid and petrol seem to cause the most damage, and if not cleaned away quickly can penetrate under the paint surface and damage a larger area. I did repaint the floor one winter and the bedroom extension on top of the garage took on a very painty smell which was not popular.

Expect to pay: £20 to £30 for a 2.5-litre tin. Larger quantities can be more cost-effective. A 2.5-litre tin should cover 10 to 15sq m subject to porosity of surface.
Pros: Simple to paint, low cost. Can improve lighting in workshop.
Cons: Can wear, repainting needed periodically depending on level of activity undertaken. Odour can be strong – care needed if internal garage.

VINYL FLOORING

Often referred to as 'lino', an abbreviation of linoleum, modern vinyl flooring is probably more associated with kitchens and bathrooms and not considered a natural choice for a garage floor. Domestic vinyl flooring is commonly cushioned for such environments, and sold off the roll. Such cushioned variants are less suitable for a garage as they are more susceptible to damage from any dropped sharp items. However, tougher versions, often referred to as Altro flooring are available, and commonly used in kitchens and hospitals. These have grit particles bonded in to help prevent slipping, and are extremely tough. A variety of colours are available, and if extra wide widths are needed sections can be joined by 'welding' them together using a infill roll and heat bonding them together. Such techniques require specialist equipment, so fitting is best entrusted to flooring suppliers.

Another useful fitting technique is rolling the flooring up the wall of the room being covered to form a skirting with a smooth curve. This can then be easily mopped/wiped to keep areas clean. If having such a system professionally installed there may be a need to screed your whole garage floor ahead of fitting, as joint areas need to be level. If self-fitting domestic vinyl off the roll this may not be necessary. DIY rolls of vinyl flooring are typically available in 2m to 2.5m wide rolls. Vinyl self-adhesive floor tiles are also available, though unless a garage floor is very smooth these may not be the best approach for a garage. Joints may also be susceptible to liquid ingress.

I would consider a cheap loose lay vinyl in a single garage if I needed a quick and easy solution.

Expect to pay: £15 to £20 per sq m for heavy duty Altro flooring. £150 to £250 for budget rolls (2.25m wide x 6m long). Fitting of Altro flooring can vary greatly according to supplier – single garage quotes obtained varied from £700 to £1200 all in.
Pros: Smooth continuous finish achieved which is easy to clean. Tough Altro style flooring can be repaired with sections 'welded' in if required.
Cons: Budget vinyl easily damaged. Base surface may need preparation. Professional installation recommended for tough Altro style flooring. Can be expensive.

PORCELAIN/CERAMIC TILES

A solution that is becoming more popular in garage environments is ceramic/porcelain tiles. Often seen in car showrooms, this is an option the average DIYer can contemplate. A fear may be the resilience of the tiles, and risk of damage, but the correct tile fitted well is very strong. Trolley jacks and axle stands can be used without difficulty on a well laid floor. Tiles need to be set on a solid bed of adhesive, ensuring there are no air pockets below, and ensuring all tiles are evenly laid, with no edges sitting raised.

Tile choice is also key. Tiles are commonly graded in terms of density, moisture resistance and skid resistance – often referred to as the PEI scale. Tiles scoring 4 to 5 are typically regarded as suitable for high traffic areas such as shopping centres, and would be ideal for a garage environment. However, these can still be damaged – dropping a starter motor on a tiled floor is certainly going to do some damage. A porcelain tile is more suited than a ceramic tile, as these are typically stronger, with an even colour/composition throughout the tile. A ceramic tile has a surface glaze/finish, so if chipped will show as a different colour below the surface.

Tiles are impervious to most fluids and can be easily wiped clean. In terms of wear it is only the grout that may show signs of dirt/aging.

Prices for tiles can vary greatly, but bargains can be found – particularly if you are happy to take end-of-line tiles. When costing a job, do make sure you consider your adhesives and grout too – these can often raise the cost significantly.

Tiling a single garage may take one to two days and grouting will need to follow once the adhesive has cured. Rapid set tile adhesives are available if time is an issue. Most tiles can be cut simply with standard tile cutters. Tougher frost resistant tiles can be more problematic, and may need power tools – I have found a masonary disc on an angle grinder to be useful for cuts.

A tiled floor is simplest to lay on a raw concrete floor as the adhesive can bond to this effectively. If concrete floors have been sealed or painted previously, additional steps may be needed to provide the key needed, or specialist adhesives. In these cases expert advice should be sought.

If you are not confident in fitting floor tiles, many local tilers will be happy to quote on a fit-only basis if you have sourced tiles, though most would want to use their own preferred adhesives. Care is needed at entrance ways to protect the edge of the first tile – a concrete ramp or protective trim is recommended.

Having tiled many large floor areas myself, I was very interested in this solution. However, due to previous coatings on my floor I felt tile adhesives may not bond successfully without investing in expensive professional solutions.

Expect to pay: £15 to £40 per sq m depending on tile choice. Adhesives £20 for 25kg – typical coverage 4 to 6sq m. Fitting costs vary, but expect to pay £20 to £35 per sq m depending on the complexity of the job. Larger tiles cost less to fit, avoid fancy patterns to keep cuts and cost down.
Pros: Great finish, long lasting and easy to clean.
Cons: Cost can be significant if not self-fitting. If not fitted correctly, can be more susceptible to damage. Can be damaged by dropping heavy items.

Different colours allow you to get creative!

PLASTIC FLOOR TILES

These are fast becoming a popular home improvement, and seen as a cost effective means of improving a garage finish. There are many types on the market, typically PVC or tough vinyl construction, interlocking in either a hidden fixing format, or a more traditional 'jigsaw' style. Bearing in mind the state of my existing floor, I felt this was the best approach for my own workshop. That said, I was still surprised at the variety on the market, and the differing approaches of manufacturers. Given the large choice available I sought samples from a number of suppliers.

Colour choice is significant in this market, with many manufacturers offering wide selections of colours and even custom palettes. Such tiles are made from virgin plastic materials, and therefore bear a higher cost.

Many manufacturers offer recyled versions, which are typically grey or black, and less glossy. These can be significantly cheaper, and equally hard wearing. Plastic floor tiles are typically laid as a 'floating floor' across your existing floor finish. This is a similar process to laying a laminate floor at home, and requires a gap (typically 5 to 10mm) around the edge of the floor to allow for expansion. Certain tiles also require an underlay. Edge pieces are commonly available for entrances, and some manufacturers also offer matching skirting profiles to cover expansion gaps.

The key factors to consider when choosing a plastic floor tile are:
- Quality of your base floor – will the tile cope with irregularities?
- Ability to deal with water spills – does the tile allow drainage? A wet car can bring a lot of water into a workshop. Do you wash parts/cars in the area?
- Is the area to be a garage or a workshop? Lifting equipment, engine trolleys and tool chests can put more pressure on a floor tile than a large car tyre – will they cope?

A tiled floor looks great, and ceramic tiles are surprisingly resistant to damage.

In the garage

Various tile styles are available. These have a flush interlocking fit.

It's recommended that you start with an L-shape along two walls.

Big Bug installation in progress. A job an amateur can easily undertake.

- Is colour important? Choose something to suit the area, not clash with existing finishes or your car, and help light balance.
- Do you want matching skirting trim? Can be a nice finish but adds to cost of installation
- Do you anticipate having to replace tiles occasionally through damage etc? Can individual tiles be lifted or would you need to remove a whole section?

Pros: Easy to self-fit, damaged tiles can usually be swapped out if needed. Easy to clean.
Cons: Some can be noisy underfoot. Not all resistant to heavy loads and lifting equipment. Edge expansion gap can be a dust trap if not covered with skirting.

As garage tiles were my chosen method, I sourced samples from a number of suppliers:

Dynotile: Dynotile supplies its standard floor tile in a wide variety of colours, with ten standard options in stock. They're 305mm by 305mm and connected by tapping joints with a rubber mallet. Installation is recommended on a flat surface, with an underlay used to dampen any noise from the tiles moving on their multiple touch points. Tiles can be cut using basic power or hand tools. The tiles fit together well, and sit on a honeycomb design with sections of different height supports allowing water to pass under tiles. The tiles are quite solid, so can rock slightly generating noise – hence the use of an underlay. When using jacks and axle stands, Dynotile advises that in many cases these are fine but because there are so many varieties of jack or hoist, and weights of vehicle, they would always recommend a board is placed underneath to spread the concentrated load. The same applies to anything that transfers large amounts of weight through small areas of contact with the floor.

Typical installation cost for a single garage (8ft by 16ft):
£369 plus shipping

Typical installation cost for a double garage (16ft by 16ft):
£699 plus shipping

Racedeck Flooring: Racedeck Flooring supplies a variety of styles and colours in two main sizes – 12in and 18in. I have focused on solid versions of these, available in gloss and matt finishes, though they also supply perforated versions (Free-Flow) which can be useful in areas where more drainage is required. Installation is achieved by either simply walking the joints or a rubber mallet could be used (though shouldn't be necessary), plus cutting tools for edges where required. The base contacts in many locations whilst also allowing fluids to pass through. No underlay is required and the design is stated as tough, supporting four post lifts and lifting equipment. High gloss versions are more expensive than matt finish and 18in versions are more affordable, and also take less time to install! The 12in and 18in squares can be laid together giving even more pattern options.

Typical installation cost for a single garage (8ft by 16ft):
£265 to £384 plus VAT and shipping.
Typical installation cost for a double garage (16ft by 16ft):
£530 to £768 plus VAT and shipping.

Mototile: Mototile supplies a variety of options. The seamless range is offered in a wide range of colours and simply installed using a rubber mallet and cutting tools for edges where required. The joints are hidden, hence the seamless name. This tile is described as ideal for residential use, ie a garage used to house a kit car, but not necessarily to be exposed to heavy lifting equipment and construction. The colour range is impressive. Matching skirting trim is also available.

Mototile Motolock is a heavier duty version capable of taking high loadings including forklift trucks. If planning a full build this would be the sensible choice from their range. The tiles fit together in a jigsaw style, and are available either in recycled grey/black finishes or in six colour finishes. Recycled tiles are cheaper. The underside design allows for water escape and ventilation.

Mototile Motomat is a heavy duty jigsaw style tile noticeably thicker than other variants, at 12mm. It is offered in a much smaller range of colours, with only yellow (virgin material) or a recycled black finish. Being thick, it is recommended for areas needing noise or thermal insulation.

Typical installation cost (based on a 3m by 5m garage)
Mototile Seamless – under £400
Motomat – under £300
Motolock Recycled – under £320
Motolock Virgin – under £450

Big Dug: Big Dug supplied samples of its heavy duty interlocking black floor tiles. These are another larger tile made from a heavy rubber type material. They interlock in a jigsaw style and are available in black and grey only. The grey colour is more expensive, and currently available in limited numbers. Edge profiles are available in yellow, which can be useful when marking out edges and entrances. The thickness offers sound deadening, and also insulation properties. Ramped edges are also available in both the male and female profiles. Due to the jigsaw type design, these are easy to fit, though important to note the underside design does not allow for water egress. Cost per tile decreases as more are purchased (lower rate for over 75 tiles, further saving if ordering over 150).

Typical installation costs (based on 3m by 5m garage)
£200 by £300 plus VAT depending on edge requirements and colour choice

INSTALLATION

Having walked over and worked on all of the samples obtained for this article over a few weeks, I have to say I was impressed with them all. I was concerned that when wet or dusty they may be slippery, but no falls so far! All seemed easy to clean, and despite many parts being moved over them and stored on them no damage was sustained.

For my installation, I decided to go for the Racedeck XL tiles. My decision was based on wanting a tile that stated it could withstand jacks/axle stands, offering some colour choice and, with the larger size available, would be a quicker installation.

Here is my overview of the installation. Whichever manufacturer you choose, all have excellent installation guides online, and many have floor designer tools you can use to help you plan your work of art! I kept it simple with mine, and chose the classic chequered design in black and graphite. I was tempted by racing livery colours, but I didn't want to create a future colour clash with future project cars! Also, I was keen to choose a colour spectrum that would hide the dirt.

The Racedeck installation guide advises starting with a line behind your garage door, and then down one edge of your garage, effectively laying an L-shape to start. My garage features a number of brick buttresses down each edge as well as steps, plumbing and power socket trunking so this wasn't possible. I still started behind the garage door – and the most important point to note here is ensuring the orientation of your tile is correct (pegs and loops) to connect to your ramp edging.

I had marked on the floor where my ramp edging was to fit, and used this as a reference point. I also measured to the centre of my opening, as I wanted to ensure the design was symmetrical. Fortunately for me, this resulted in the offcut from the cut on one leading edge being sufficient to fill the gap on the other side, so reducing wastage. When using the Racedeck planner, I had found the 18in tiles would give me the simplest installation in terms of cuts. This is a key area to consider with your own installation in terms of cost of wastage, but also work involved cutting.

The Racedeck guide is also very clear that the best approach is to empty the whole garage first. However with the British climate to work against, and two roofless kit cars I didn't have that luxury. I actually started behind the garage door with just the buggy removed, and laid a complete line of alternating tiles across the entrance. I then worked back in rows, leaving those tiles against the edge for later cuts. In around 40 minutes I had laid over half the garage and was able to wheel the buggy back in from the rain! The next evening I repeated the process for the other side of the garage – this time leaving the Nova shell in the rain.

Again, around 40 minutes allowed me plenty of time to lay the remaining whole tiles

Starting point for the installation in James's garage was at the door end of the garage. Ramps in, too.

Initial section laid. James couldn't entirely clear out his garage because of the weather!

A closer look. The joins in the Racedeck tiling are seamless once installed.

Assistant Martha proved that installation is child's play!

You can either 'tread' the tiles together, or tap them with a mallet.

Gaps will be filled when summer weather warms and expands the tiles to their maximum.

Wall paint extends to the floor to help hide the expansion gap.

Fine cuts best done with a jigsaw. Don't go too slowly, though, or it will 'weld' itself together.

Cutting with power tools gets dusty, so it's best to work outside!

In the garage

and leave myself just the cuts to finish. I must stress this is by no means the best way to hit the job, but does show how simple it is even in trying circumstances! I found the clipping together of the tiles very simple, and as per the installation video walking on the joints when the tiles are laid together simply locks them in place. As I had my spare tiles laid out in piles around the garage I often found myself on my knees and it was simpler to tap together with a rubber mallet, rather than get back up – but that was my preference. This was also necessary in some of the awkward areas at edges and under my work bench where I couldn't walk the joint.

The following weekend, I was fortunate to have a rain-free day, so all cars, tool chests, engines and jacks were fully removed from the garage to allow me to complete the cuts. As previously mentioned, I needed to cut tiles for both edges of the garage, due to the design, and very irregular brickwork. I also needed to trim the back wall edges under my workbench. These cuts were less visible so I didn't have to be so precise.

My garage doesn't get a lot of direct sunlight, but can warm up a fair bit in summer, so I am waiting to see if my expansion gap is sufficient. I am not going to fit skirting to cover this yet, though this may follow if I find it becomes a dust trap. If you have a more regular garage you will certainly have fewer cuts than I did. With the workbench legs to contend with, two concrete steps and three uneven edges I needed to make around 40 cuts. Despite this, it was only a further two hours all in to complete all edging and put everything

Circular saw good for straight cuts.

Working around workshop legs and step took time.

Lots of cutting out to be done here!

Five hours saw the job complete.

back in its place. For small simple cuts to the ramps, I found a sharp Stanley knife did the trick. I also used this for cutting around bench legs. Full tile straight cuts were achieved using a circular saw as I have one, but again would be easy enough with a good handsaw. For L-shape cuts and awkward sections I did use a jigsaw. I found with this the key was to move fast with a fine sharp blade. If you stay in the same place for too long, as I did when trying to cut a curved section to slot around pipes, the plastic can melt and solidify in the cut. This is easily resolved with a quick snap and sandpaper though. Cutting using any saw does generate a fair bit of mess, so outside work is ideal.

First impressions once fitted – my daughter thinks it is wonderful to scooter on! I find the garage lighter, and when I dropped a container of engine bolts on the floor they were certainly easier to find – and less gritty! There is a slightly different sensation under foot, and audibly. Hopefully it will warm up the workshop slightly too.

Overall, the job took me about five hours. Had I done it in one session with help from someone over three years old, or had fewer cuts around pipework and buttresses, it would inevitably have been quicker. Every time I look at it now I do wonder why I didn't do it sooner – much more satisfying than smelly paint!

Cars brought back in. Looks great!

Impact 'Drivers

John Dickens shows you how to use an impact screwdriver when stripping down old donor components.

A typical Impact screwdriver kit (now over 40 years old).

It still removes rusty brake drum screws easily and without damage.

The tool and the accessories are made entirely from steel.

The hexagon drive bits are hardened and tempered.

This is a ½in square drive impact screwdriver.

An impact screwdriver is a tool which, when struck with a hammer, converts the energy of the blow into a high torque rotational force. Most people will be aware that it is sometimes possible to 'shock' a threaded fastener loose with a sudden blow even when a large force, applied gradually, has failed to loosen the nut or bolt. This is the principle on which all impact tools work.

I first came across impact screwdrivers in the early 1960s when people began to work on the recently imported Japanese motorcycles. Their engine casings were usually held on by Philips head machine screws made from relatively soft steel, and it was difficult to remove these without damaging the screw head as the tapered form of a Philips screw head tends to cause the screwdriver to lift out and slip. Impact screwdrivers overcame this problem perfectly since at exactly the same time as the rotational force occurs, the

In the garage

This smaller 3/8in drive tool is more versatile as it is a more common size.

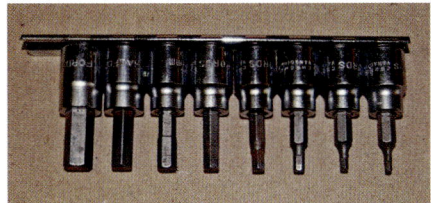
These Allen screw sockets will fit the smaller drive system.

The socket simply replaces the supplied chuck.

The increasingly popular Torx sockets can also be used.

This VW cheese head tinware screw has not moved for some time.

The chuck and screwdriver bit are pushed on to the tool.

With the chuck held still the body is rotated fully anti-clockwise.

The tool is struck firmly with a large hammer to rotate the bit.

Once the screw has loosened, an ordinary screwdriver completes the job.

hammer blow also causes a large downward force on the screw preventing any lift or slip.

I have owned my current impact screwdriver since I bought my first Japanese motorcycle in 1975 and it is still in use on my own vehicles.

CONSTRUCTION
Impact drivers need to be strong and sturdy as they are designed to be hit hard with a hammer. They are normally made entirely from steel but some have a rubber or plastic grip over the main body. They are usually supplied with an assortment of straight and Philips screwdriver bits with hexagonal bodies which fit into the chuck on the tool. This chuck is also removable and normally utilises the same standard square drive found in socket sets.

My particular example uses a 3/8in square drive which increases its versatility greatly as it can also be used to drive Allen sockets, Torx sockets or even standard hex sockets should the need arise.

The direction of rotation is normally reversible so that they can be used to unscrew left-hand threads too.

IN USE
Decide which bit or tool is most suitable for the fastener causing the problem. Push the chosen socket or bit firmly into place until it clicks. These pieces are normally retained by spring clips. The next step is to select the correct direction of rotation. Hold the chuck firmly in one hand and turn the body of the tool as far as possible in the required direction of rotation. This would be anti-clockwise to undo a normal RH threaded screw.

Engage the screwdriver bit with the screw and turn the tool anticlockwise to take up any play. Strike the tool firmly once with a large hammer (a copper hammer is ideal for this). One firm strike is far more effective that a number of gentle taps.

If the screw does not move re-engage the tool, take up the free play and try again. Once the screw begins to move, a normal screwdriver can be used to fully remove it.

Obviously care should be exercised on more delicate materials such as alloy castings or thin panels and bear in mind that in cases where the fastener is really seized solid it is still possible to damage the head of the fastener or even break it off with the high torque available.

Additionally, although it is possible, I would never use an impact screwdriver to tighten a screw as it is very difficult to control the torque applied and it would be easy to strip a thread.

SELECTION
Most tool manufacturers produce their own version of this tool so they are readily available. Prices range from around £5 to £20 depending on the quality of the tool and the accessories supplied. As with many special tools this is not something that you will use frequently, but on those occasions when it is needed you will find it to be well worth the purchase price.

Airless Spray Gun

John Dickens explains the principle behind airless spray guns before putting theory into practice. Why not give it a go with your own car?

A typical air-fed spray gun.

Conventional paint spraying equipment uses compressed air, normally supplied from an electrically driven piston type compressor, to atomise the paint and distribute it onto the work surface via a spray gun. The paint is stored in a container integral with the spray gun and (in a siphon type gun) is drawn into the air stream by the Venturi effect of the airflow. The paint is then mixed with the high pressure air, forced out of a small nozzle and blown onto the object being painted.

Spray guns of this type are very versatile, having adjustments for airflow, paint flow and spray pattern. Typically this process may require a compressor delivering 10 to 12 cubic feet of air per minute (cfm) at a pressure of 50 to 80psi. A relatively recent innovation is to use a larger volume of air supplied at a lower pressure. These High Volume Low Pressure (HVLP) guns typically use 20cfm of air at a pressure of around 25psi. They produce less overspray and bounce-back than high pressure guns, so they generate less waste and mess. Hobby versions of these HVLP systems use turbines rather than compressors. These generate very low pressures but move very large volumes of air. I have used a Wagner W800 unit of this type for years and have produced some very good results with it.

An alternative spraying system uses no air at all. It works by pumping the paint through

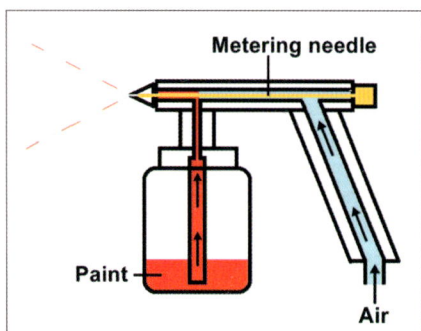

Paint is drawn from the reservoir and mixed with the compressed air.

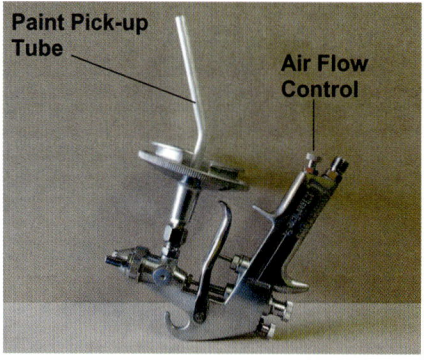

This control regulates the total air flow.

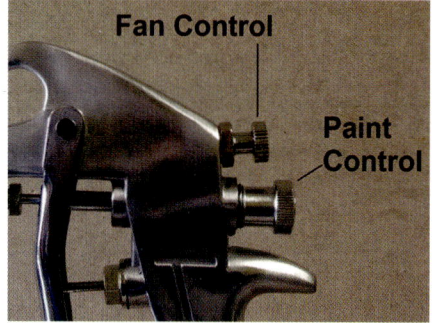

The volume of paint and the shape of the fan can be varied.

90 *Modify, Improve & Upgrade Your Kit Car*

In the garage

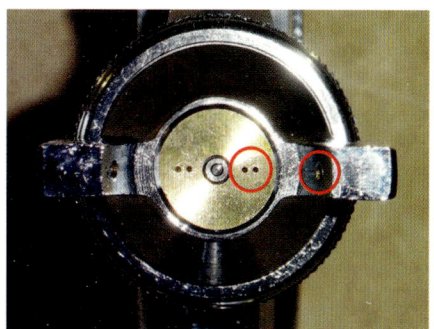
Air coming from these holes controls the shape of the fan.

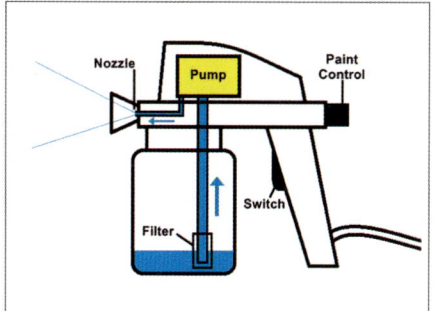
The airless spray gun is a completely self-contained unit.

The Titan SF75 is a typical example of the airless paint sprayer.

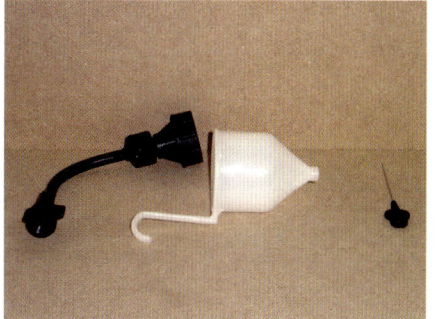
A viscosity cup, an angled nozzle and a jet cleaner are included.

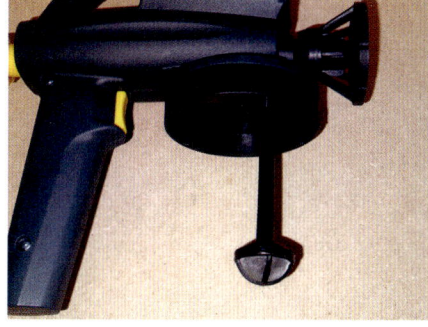
The filter prevents any solid material entering the pump.

For this test, John used an oil based polyurethane paint.

the nozzle at very high pressure causing it to atomise into fine particles and spray onto the work. Larger systems store the paint and pump in a separate unit, connected to the spray gun by a flexible hose, but the type of gun on test here is of the self-contained hand-held type which incorporates the paint reservoir, the high pressure pump, the mixing nozzle and the controls in a single unit.

The operation is very simple. The pump is an electrically driven piston type component which draws the paint from the reservoir and pumps it, at very high pressure, (150bar/2250psi) to the nozzle where it atomises into a fine spray. The only external connection required is to the mains electrical supply. The only adjustment provided is a control to vary the volume of paint being sprayed.

The gun used for this test is a Titan SF75 80 watt unit. This particular model has now been discontinued but many other similar systems are available. It is very much at the budget end of the market costing less than £20 from Screwfix a year or so ago. Prices for this type of spray gun typically range from £20 to £60.

Airless paint sprayers of this type are normally intended for interior decorating.

White spirit is a suitable thinner for oil based paint.

They are an alternative to using a roller or brush on large areas and, as such, are not designed to produce the type of fine finish normally associated with automotive spraying systems. I decided to use the gun to paint the remaining pieces of tinware for my Beetle engine, as surface finish is not an issue on these components.

The gun itself is a neat compact and lightweight unit with all the external components being moulded in plastic. Also included in the box is an extension nozzle for painting horizontal surfaces, a viscosity cup and a needle to clean the nozzle should it become blocked.

Viscosity cup is used to thin paint to correct level.

Removing the paint container reveals the filter on the paint pickup pipe. All these airless guns need filtered paint as their pump tolerances are very tight and even small solid particles can cause damage. Since these guns are intended for home decorating, not all of them are suitable for use with cellulose or synthetic lacquers. The instructions for the Titan gun did not mention these paint types so I decided to play safe and use an oil based polyurethane paint instead. In common with most paints intended for interior work this turned out to be a thixotropic 'gel' paint but, since it would need to be thinned for spraying anyway, this was not a problem.

Modify, Improve & Upgrade Your Kit Car

Stirring the gel paint disperses the pigment evenly.

It also allows the thixotropic paint to be poured.

Some paint is poured into a separate container.

A small amount of white spirit is added to the paint.

Mixture needs thorough stirring to disperse thinners.

This should take 18 to 22 seconds to drain fully.

Thinned paint is transferred to the paint container.

Spray gun is then assembled and is ready to use.

Paint flow adjusted to produce best spray pattern.

PAINT PREPARATION

White spirit or turpentine are suitable thinners for oil based paints and are easily available. All paints need to be thinned for spray application, even those designed specifically for automotive use. Two pack paints are thinned by mixing the paint and the hardener in a specific ratio. Cellulose and oil based paints are thinned until they have the correct viscosity for the paint type and the application system in use. A viscosity cup is used to achieve the required paint thickness.

All paints should be thoroughly stirred before use to evenly distribute the pigment but in this case stirring the paint also makes it liquid enough to be poured into a separate container for thinning. The thinner is added a little at a time and stirred in thoroughly, especially during the initial stages when the paint is still thick. To check the viscosity of the paint the cup is first immersed in the liquid until it is full then lifted out and timed as the paint drains through the calibrated hole. The instructions for my gun specified a time of 18 to 22 seconds for oil based paints. I would estimate that I had to add about 25per cent white spirit to the paint to achieve this.

Once thinned to the correct viscosity the paint container can be filled and attached to the gun. There must be enough paint to keep the pickup tube fully immersed as the pump cannot be allowed to run dry. The gun must also be kept fairly level in use to avoid exposing the paint pickup tube.

SETTING UP

Before use, the spray gun must be set up to give the desired spray pattern. There is only one adjusting screw on this gun so the adjustment is very simple. The nozzle is pointed at a vertical surface (I have a garage wall reserved for this) the trigger is pressed and the screw is adjusted until the correct spray pattern is achieved. I found that the best spray pattern was produced with the paint flow set to maximum. Below this setting the gun just sputtered and blew out large droplets of paint.

IN USE

I had prepared the VW ducting panels by flatting the 'factory primer' with a Scotchbrite pad. (Why is 'factory primer' so fragile in use but so tenacious when you want to remove

In the garage

Tinware has been flatted to provide a key for paint.

Gun is held well away from the work to avoid runs.

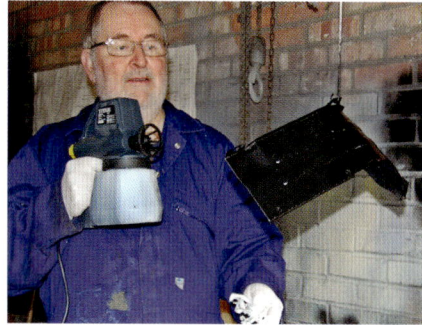
Avoid tilting the gun. It may allow it to suck in air.

A lot of the paint ended up on the garage floor.

The paint container is removed first for cleaning.

Unused paint is poured out. Useable for a day or so.

Paper towels used to clean up the remaining paint.

White spirit flushES out remaining traces of paint.

Solvent is swirled around container to clean it.

it?) The panels were then suspended from the garage roof for spraying.

I tried various spraying techniques and held the gun at various distances from the work piece but, to be honest, I was not particularly impressed with the performance of this spray gun. Each time the trigger was pressed it seemed as if the pump had emptied and the gun initially produced large droplets instead of a spray. Even after the paint flow was fully established, atomisation was poor and the spray pattern contained droplets of paint rather than a fine spray.

Since the paint flow had to be set to maximum to optimise the spray pattern, the gun needed to be held a fair distance away from the work to prevent runs and sags. This meant that most of the larger paint droplets simply fell onto the garage floor. Eventually I got some paint on the panels but the wet layer was far from flat and also contained a large number of tiny bubbles. Had the paint been cellulose or some other quick drying coating, the surface finish would have been very poor. Fortunately, the slow drying oil based paint allowed the orange peel effect to flow out and most of the air bubbles to dissipate before it began to harden. There were still a number of runs though, in spite of my care. Had I been careless with the paint preparation I could have excused the poor performance of the gun, but I spent a lot of time thinning the paint to the exact viscosity specified in the instructions so I can only presume that this is as good as this particular spray gun gets.

AFTER USE
All spray guns must be thoroughly cleaned out after use. Hardened paint residues in the fine drillings and passages can make any gun unusable. The Titan gun was cleaned in much the same way as any other. After use, the paint container was removed from the gun and unused paint was emptied out. Any paint remaining in the container was cleaned out using tissues or paper towel and the container was flushed out with white spirit. Clean white spirit was then pumped through the gun and sprayed from the nozzle to clean out the internals. The exterior surfaces were also wiped clean using fresh white spirit. The nozzle was unscrewed and thoroughly flushed out and the small sprung piston

Clean thinners blown through gun to clean internals.

White spirit will also clean the outside of spray gun.

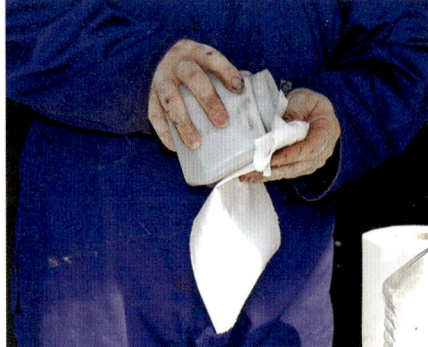
The cleaner the better.

Better than brushed; acceptable at best for a sprayed.

The nozzle needs to be unscrewed for cleaning.

All traces of paint need to be cleaned out.

The internal components also need to be removed.

These also need to be scrupulously clean.

The stylus is used to clean out the nozzle jet.

behind the nozzle was also removed and cleaned. Finally the nozzle was cleaned out using the tool provided.

CONCLUSION

I applied three coats of the polyurethane paint using the Titan SF75 and allowed them to dry overnight. The finished result, whilst being reasonable for an engine component, is not what I would consider acceptable for a spray painted finish. I have used other airless spray guns in the past and I know that they are capable of producing much better results than this. I also know that Titan produce a large number of different airless spray units for professional use, so I suspect that the problem lies with the budget nature of this particular piece of equipment.

Hand-held airless paint sprayers of the type tested here are not designed for spraying or re-spraying car bodywork. The surface finish they can achieve will never be as good as that produced by a turbine or compressor driven air fed spray gun. They are, however, ideal for spraying chassis, suspension and external engine components and can apply paint faster and produce a far better finish than could be obtained by brush painting these same components. They can also be used to apply PVA release agent, household paints and wood preservatives should the need arise.

My advice, as always, would be to buy a good quality unit from a reputable manufacturer but check, before buying, that it is suitable for use with cellulose based paints. Incompatibility with these paints will severely limit the usefulness of the spray gun for automotive use. In terms of price even the more expensive versions are still quite reasonable at around £60. When you consider that some of the larger aerosol paint cans are around £12 each, a good quality airless spray gun may prove to be a good investment in the long run.

Interior Trim

Ian Stent looks at the different trim materials you'll find in a typical kit car cockpit, and outlines the options you should consider when designing your own trim.

Vinyl comes in a variety of different grains.

Vinyl can have a wide variety of different finishes.

The centre panel here is a carbon effect vinyl.

Do you understand what materials are available for use in a typical kit car interior? How they can be joined or finished? How they can be combined? And how they should be attached to panels? There are now myriad materials and finishes suitable for use in your car and, with the help of Interiors Seating in Nottingham, CKC has been on a crash course to outline your options so that you can achieve the very best finish possible for your pride and joy.

There are three main covering materials you'll be considering for your car, unless it's a stripped out track day specialist with bare panels. These are vinyl, leather and carpet. In addition, suede or fake suede (such as Alacantara) may come in for consideration and beneath vinyl or leather you may be considering a sub material such as foam. So we are going to first look at the materials, and then ways or finishing and joining them...

VINYL

Affordability is one reason to choose vinyl instead of leather, but there are several other reasons why you might choose this manmade material over its natural forbear.

Ease of use – Vinyl comes in a uniform roll and so can be ordered by the metre length and you can utilise as much of the material as possible on your trim panels. Leather obviously comes as a single hide, with differing finishes and thicknesses and imperfections across the product.

Practicality – In an open car, getting vinyl trim wet in a shower isn't a major headache, but you'll want to avoid it on leather.

Visual finishes – As well as the obvious different leather effect surface finishes and different colours, vinyl can also be bought with a variety of other finishes, such as a carbon look, metallic colours, perforated and much, much more.

As you'd expect, there are different qualities of vinyl, and the backing material can vary too. Generally speaking, better quality vinyl is thicker, and you can see the difference when looking at the edge of the vinyl. Thicker vinyl is less likely to tear and

Perforating vinyl or leather can look great.

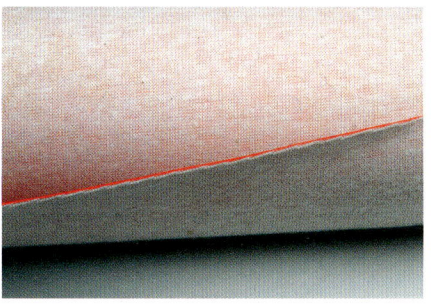
This is good quality vinyl. The vinyl layer is thick.

Some backing allows the material to slide over foam.

Leather is supplied as a whole hide.

Leather comes in different colours and grains.

Suede is the 'untreated' leather.

Alcantara is lighter in weight and comes in a roll.

stitching is less likely to pull through but, for the amateur, thicker vinyl is less flexible, which can make it more difficult to shape around curves. That said, Interiors Seating points out that it's possible to make thicker vinyl more flexible with gentle heating.

Some backing is more fabric-like than others, and this can make the material move over any foam or sub material more easily, which is generally advantageous.

When it comes to price, most rolls of vinyl are typically around 1.4 metres wide and sold by the metre length. Expect to pay anywhere between £15 per metre for a standard grain finish to over £40 per metre for special effects.

LEATHER

As already mentioned leather is obviously a natural substance and it's typically sold by the hide. As with everything, prices can vary, and Interiors Seating produced the remains of one hide it has used recently... at £700 for a single hide! A more usual price would be around £150.

As already mentioned, the leather across the hide will vary in thickness and flexibility, so certain areas are more suited to making gear lever gaiters, while others will suit seats or larger trim panels. The visible surface of leather is still usually a treated surface which not only gives it a colour, but also a variety of different grained finishes. You should think carefully about the type of visual finish you want, and see different samples before you make your choice. If you know how much leather you'll need, then it's probably sensible to order multiple skins at once, to ensure a consistency of colour and treatment.

There's no doubt that there's something special about the feel and smell of leather... it certainly adds a touch of quality to any interior. Having said that, there are some increasingly convincing looking vinyl products around.

SUEDE REAL/FAKE

Suede is effectively the untreated leather, which is what you see on the underside of a treated piece of leather. It's an expensive product because, since it's largely untreated, it needs to be carefully chosen in order to be free of visual imperfections. However, used as part of a trim package, size and cost can be controlled.

There are a number of fake suedes available and perhaps the most well known is Alcantara. Very popular in the late '80s and '90s, Interiors Seating is seeing less use of it more recently.

Alcantara is considerably lighter and thinner than suede, but the surface finish is terrific. Unlike its natural inspiration, Alcantara can be ordered by the metre and maximum usage can be made of a material that is utterly consistent across its surface. It can be ordered in a wide variety of colours and, like leather and vinyl can be perforated.

Ideal for use as a non-reflective dash material, Interiors Seating does not recommend it for use in high wearing areas, such as seats, since the surface can become flattened and cannot be easily recovered.

CARPET

Carpet is the other main material that you'll be dealing with in a car cockpit. Before going any further, it's worth pointing out that domestic carpet is not suitable for automotive use, where it will quickly rot and is not designed for the job.

At its most basic, you will be opting between synthetic and woollen automotive carpets. Almost all will have a rubberised backing which binds the carpet tufts together and does not rot. Hessian backing is another more traditional finish sometimes found on woollen carpets, but which is probably not relevant for the modern kit car builder. Carpet quality can vary enormously, and it's

Carpet usually either synthetic (black) or wool (red).

Most automotive carpet is rubber backed.

Colour choice is quite limited.

Trim

Interiors Seating's Ash demonstrates binding.

worth getting along to a show to look at and feel different carpets from the suppliers. Alternatively, most suppliers will be happy to send you a small sample in the post.

Cost may be a major deciding factor in the type of carpet you choose, with a woollen carpet often being double the price of the equivalent length of synthetic carpet. If you are only trimming a small area, then cost may be less of an issue, and the editor recently went for a wool carpet when retrimming his Cyclone. Generally speaking, Interiors Seating finds most people opt for a good quality synthetic carpet.

The rubber backing works well with adhesives and, because it's rot-free, it's not vital to use any form of underlay and the carpet can be glue directly to internal panels.

EDGING/BINDING

Vinyl/leather – Vinyl and leather is often wrapped around a panel, so the fixing is concealed around the back. Typically, this will be via glue or, if on a thicker wood panel, perhaps stapled. But there are times when two pieces of material need to be stitched together, perhaps in order to cover a more complex dash moulding. See the following section on Visual Effects for the various stitching options.

Carpet – If the cut edge of a carpet is left open, then even on a rubber backed carpet, tufts of material can come adrift and the edge can deteriorate. There are times when this is not a problem, such as when a carpet on the side of a centre tunnel reaches the floor and the edged floor panel overlays it, concealing and protecting the unfinished edge.

However, visible edges should be bound. This is usually achieved with the use of a strip of fabric or, more probably, a leather or vinyl strip of material which matches the material used elsewhere in the interior. 'Whipping' is another option for finishing a carpet panel, more usually found in classic cars. Binding can either finish the carpet edge, or join it to another piece of carpet to form a flexible joint, such as when carpet on a centre tunnel top, is joined to carpet used on the side of the tunnel.

TRIM EFFECTS

Piping – This is perhaps one of the most obvious additions to a vinyl or leather

A bound edge on carpet often matches vinyl/leather used elsewhere in the interior.

A whipped binding, which is more typically found in classic cars.

Binding can also join two pieces of carpet, where there is a change in surfaces.

Modify, Improve & Upgrade Your Kit Car **97**

Piping is most often used in seats.

Standard stitched join between two pieces of vinyl.

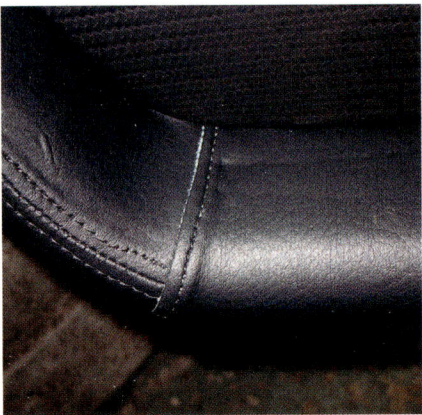
Good examples of a single top stitch (right) and double top stitch (left).

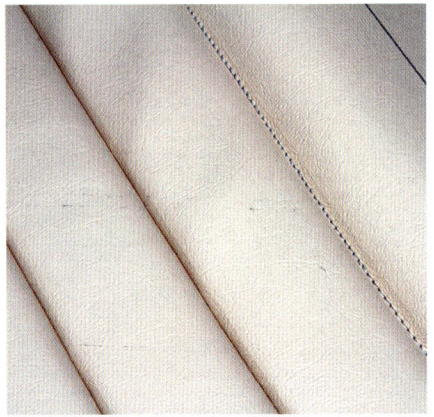
L-R: Pencil guide line, open single stitch (quilt) and examples of fluting, where stitch is hidden.

product, used to enhance its looks. Piping is most usually found in seats, but it can also be used on trim panels. Adding a contrasting colour to the interior is easily achieved with piping, which may match the exterior colour of the car. Although piping is almost exclusively used for visual effect today, the rope used inside the piping can be pulled to draw material on a seat into a certain position and hold it there under tension.

Stitching effects – When two materials are joined with a single stitch there is always some material 'beneath' the stitch which must be folded to one or both sides to avoid a raised bump under the stitched join. This can either be carefully laid to one side, or it can be stitched down on one side. This is known as a single top stitch, adding a single visible line of stitches to one side of the join. Another alternative is to fold the additional material behind the join down on either side of the join, stitching it flat on both sides... a double top stitch. The latter gives added visual interest and is often seen on upmarket interiors. Finally, it also allows the trimmer to add an additional layer of material across the back of the join which, when stitched in place, bridges the main join, adding strength to it.

Stitching doesn't just have to join two materials and can be used simply to 'draw' a design into the surface of the vinyl or

Look closely and you'll see the stitch is visible in this GD. This is quilting.

A good example of a diamond stitch quilted centre tunnel.

98 *Modify, Improve & Upgrade Your Kit Car*

TRIM

Quilting detail on a seat.

Piping detail on a seat.

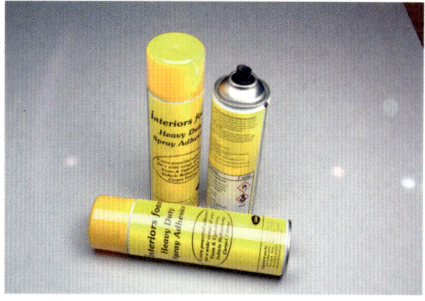
Interiors Seating's own spray adhesive.

leather. Ultimately, this could be in the shape of a logo, but it may just be to add some visual interest to an otherwise large and flat surface, such as a side panel.

Quilt/flute effect – Fluting on seats or on panels is perhaps one of the most common visual effects we may be familiar with. It can also have a practical benefit though, by stitching the surface material to the sub structure (usually foam) to stop the surface material from moving around too much. Quilting is achieved by running a stitch through the top surface, binding it to the foam beneath. Run the lines parallel and the effect is similar to piping, with a raised area in between the lines of stitches. Vitally, the single stitch line remains visible. Quilting can also be done to achieved the diamond effect you may see on some supercars and, increasingly, within kit car interior panels, such as the centre tunnel top.

The most obvious visual difference between parallel quilting and fluting, is that on a fluted panel, the stitch is no longer visible (Fig 25). More often than not, fluting is achieved once again by simply stitching the top material to a foam layer beneath it, although a cheap fluting effect can be achieved by effectively 'welding' the vinyl to the sub structure or, in a classic car, the flutes can be two stitched-together layers of material with a padding pulled through each 'tube' to form the soft raised part of the flute.

ADHESIVES
If you are trimming your own car, then you will inevitably need to glue trim into position at some stage. Adhesives are available in either a tin or spray can. We've traditionally found that the glue from a tin, applied with a plastic applicator of some description, invariably has had more long-lasting effect than sprays.

However, Interiors Trimming uses its own-branded spray adhesive (Fig 26) which it obviously finds most effective. To achieve a good result, both surfaces are first sprayed and allowed to become tack dry. A further light spray of adhesive is then given before the two surfaces are immediately joined... job done.

SEATS
A big part of Interiors Seating business is the manufacture of its seating range. Obviously, this isn't something we'd expect the home builder to become involved with, beyond specifying the colours and detail effects needed. All of its seats are steel framed for rigidity and the customer can customise any seat to almost any degree. Less seat padding, different coloured panels, pipe and stitching options... all can be customised to your requirement. As with the rest of the trim, take your time to consider the options that will make your car stand out.

SUMMARY
The whole point of this article is to outline to you the different ways in which you can trim your car. Whether it's a minimalistic sevenesque kit or a full blown Ferrari replica, there are endless ways of achieving a unique and high quality look through the use of different materials (vinyl, Alcantara etc), different effects (stitching, fluting, piping), different colours and different bindings. Heading to a kit car show is a great way to see and feel these materials, to understand how flexible they are, to see all the different surface finishes available and to talk to the suppliers to get more information. Companies such as Interiors Seating can offer different levels of service, from supplying the core materials, to making up trim kits for well known cars (such as most of the Cobra replicas) or offering a complete in-house trim service... drive in, drive out – job done.

Interiors Seating offers a wide range of different seat styles.

Modify, Improve & Upgrade Your Kit Car

Trim a panel

Tech contributor **James Horsley** outlines the process of creating a simple but effective trim cover panel, a method that can be adapted for various applications.

Most kit car interiors will require some form of basic panelling.

Storage area on James' buggy needs a cover.

MDF board marked out after using a card template.

Building or restoring a kit car takes many skills and, for many of us, this is the appeal of the process. Inevitably we all have our preferred skill sets and, certain jobs we will always pass on to the professionals. Often upholstery is one of those areas, though depending on your chosen kit, an interior can be as simple as bolting in a couple of pre-trimmed seats. However, more often, other parts of a vehicle may need trimming, even if just a small dash panel, a battery cover or basic luggage lid.

Whilst there are many small independent trimmers out there you could turn to, learning the basic skills to prepare and trim an interior panel can save a few pounds, and also be very satisfying. This guide is designed to give you a feel for the skills and approach needed. With a little practice and time, most home builders should be able to achieve a professional looking result at a modest cost.

For this article I needed to fabricate a small parcel shelf for my beach buggy (Fig 1). This panel needed to be removable for access, which does complicate the process slightly. If starting out for the first time on such a panel, my recommendation would be to begin with a small fixed panel and work up to more involved pieces. I have previously designed and fabricated small trim panels for other beach buggies, including door cards, battery enclosure covers and glove box lids. This has given me confidence to work up to trimming bench seats and even my

Rotary saw and jigsaw are both useful tools to own.

This 'fence' guides the saw for a straight cut.

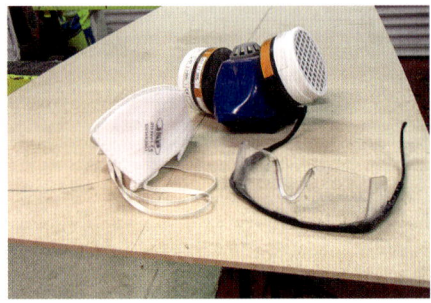

Invest in a decent face mask to avoid dust.

Trim

Jigsaw is best for creating even curves.

It's taking shape.

Panel cut to shape, but additional work needed.

Paper template used for the cut out to allow for the roll cage tubes.

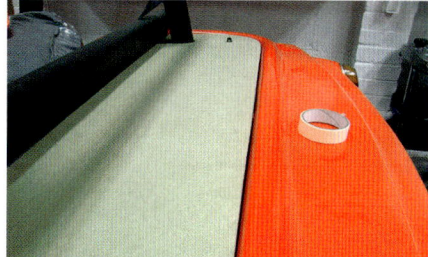
After several trial fitments, MDF panel now fits neatly in the space and around roll bars etc.

Any additionl brackets need trial fitting now, before trimming begins.

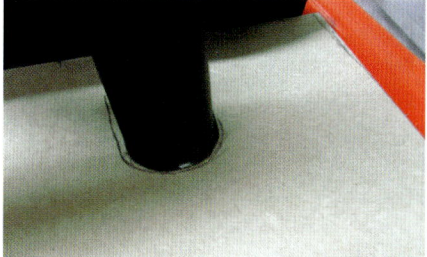
You need to allow for the thickness of the trim. The panel needs adjustment here to allow for this.

Gap around the panel looks large, but allows for material to wrap around the edges.

James varnished the panel for added weather protection.

Manual stapler works well, air compressed easier!

Wadding (or foam) gives the panel a padded finish.

Spray adhesive works well for this job...

Apply an even layer.

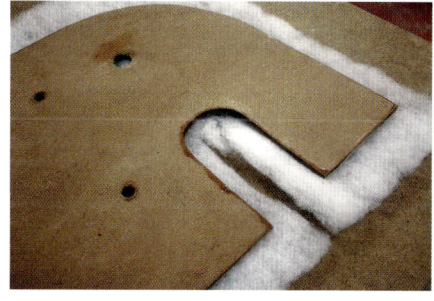
Wadding trimmed, allowing it to fold over edges.

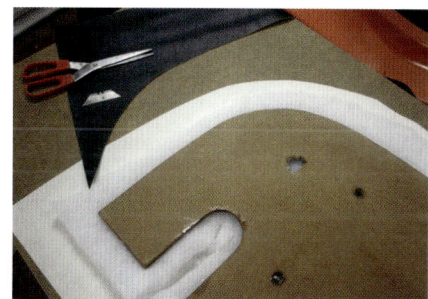
Vinyl then trimmed slightly oversize.

Modify, Improve & Upgrade Your Kit Car

daughter's trike to match it to our vehicles!

A paper/card template is always the best approach to any project. Then transfer this to your timber panel. Consideration should be given to the best timber for the project. Exterior grade plywood can offer good weather protection, but if making an internal panel this may not be a factor. Manmade fibre boards like hardboard and MDF are easier to sand and cut but can be more susceptible to moisture. All are available in varying thicknesses, which also needs to be factored, depending on the location of the panel.

Basic DIY tools like an electric jigsaw help greatly, handsaws could be used with patience (though curves would be more challenging). I choose to use a circular saw for long straight cuts, as with a fence attached smooth straight edges can be obtained. If you don't have such tools, often local timber merchants can help with basic cuts, leaving you just the detailed cuts to make.

When cutting or sanding manmade fibres in particular, care should be taken to avoid inhaling dust, so wear a suitable breathing mask.

Having trial fitted your panel several times during the trimming and preparation process, don't forget to leave sufficient gaps around the panel to allow for your chosen trim materia. If you are planning to fit any fixings to the panel you should make holes in the panel ahead of covering. Drilling through a trimmed panel is a tricky process and risks damaging the material.

As I had luggage rails and panel clips to install, I test fitted all of these on my panel ahead of the trimming process. Once completely happy with the panel I varnished it for weather protection. You may not need to do this depending on your panel location.

I chose to trim the parcel shelf in a basic black vinyl. I sourced this from a local fabric shop and it was inexpensive. Many types are available in a variety of colours via eBay, kit car shows and the bespoke trim companies you'll typically find advertising within CKC. Sometimes it's useful to talk to a supplier to understand what certain materials can offer… how stretchy they are, how suitable they might be for the panel you have in mind.

I generally steer away from fabrics with patterns, as stretching this needs to be done evenly to avoid a distorted finish.

Beneath the outer vinyl I also wrap the panel in a light foam wadding, again easily obtained from upholsterers. This helps protect the top vinyl from any splinters or imperfections on the panel, and also helps give the panel a more production padded look.

The wadding is cut slightly oversized to the panel and glued in place. Spray glue is ideal

Staples come in different lengths. Be sure they do not come through the outer face of the panel.

Vinyl loosely stapled to one side of the panel at this early stage.

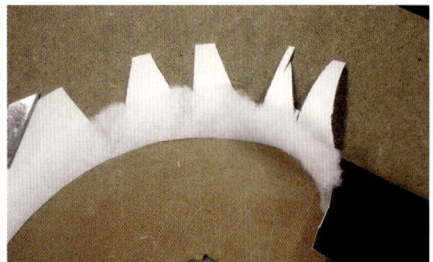
At the corners you'll need to cut the vinyl into triangular segments to make the folds neater...

...and now stapled into place.

Simple 90deg corners can be neatly folded over.

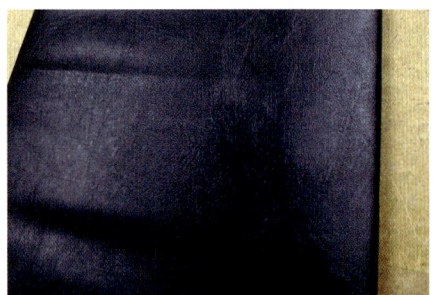
Not perfect yet. Ripples mean adjustment needed.

Exterior finish on this curved edge is just right.

More complex areas need extra care.

(29): A folded corner can look very tidy.

Trim

Surface finish now perfect. Just the brackets to attach.

Cutting out the vinyl to allow fixings. Don't cut too far or the cuts may show on the outer surface.

James also added trim to the underside to cover the MDF. Visual finish here much |less important.

The finished panel is held in place with two clips, allowing quick access to storage area below.

With care it's perfectly possible to achieve a well finished end result.

Several trimmed panels on James' previous Bugle buggy, plus matching trike seat to match!

here as it dries quickly, avoiding any delays. Once the wadding is glued into position, edges can be accurately trimmed before offering the panel to the top vinyl. The vinyl will also need to be trimmed around the panel, allowing sufficient spare material to wrap around the edges and attached to the underside.

Care is needed to stay mindful of the thickness of your panel when trimming, particularly around curves where multiple cuts are needed to stretch around these involved areas. I always start on one long edge of a panel, stapling in place every six inches or so. This helps hold the vinyl in place, but can be undone if needed. Tension can then start to be pulled by stapling the opposite side.

As this progresses, areas such as corners will need attention. Gentle curves are easiest to tackle as the material can be stretched and pulled, and then stapled – choosing your staple length wisely! I prefer to use an air-powered staple gun as less effort is needed, but a hand stapler can be used if you don't own air tools. Sharp 90deg corners can be trickier, and need parcel type folds to get a neat result. If you are good at wrapping Christmas presents you should excel here!

Care is needed to ensure you work on a clean bench or table to avoid damaging the top side of the panel while working on the underside. Also, regularly turn the panel over to check the tension is correct. If the vinyl is too loose pleats or creases will be visible.

In action at a show last year, complete with 'matching' trike!

As this panel is going to be removable, I wanted to trim the underside too. This was more to protect the timber face, so not as critical in terms of visual finish. Before doing this I used a sharp blade to pierce the vinyl where my holes were already made in the panel. I then enlarged the holes further from the top side with a sharp blade, taking care not to elongate the holes too far. I then trimmed the rear of the panel ahead of fitting the luggage rails and panel clips.

I estimate the total time spent on this panel was 4 to 5 hours, but this was spread over a number of evenings, allowing time for varnish to dry and photographing each stage. With a clear workshop and diary this is certainly a weekend project and a very rewarding way to personalise your cherished kit car.

Modify, Improve & Upgrade Your Kit Car

Poly Bushes

John Dickens takes a close look at the technology behind polyurethane bushes, and how to select the right product for your car.

Polyurethane has been used to replace rubber components in various automotive applications for over twenty years now and most of us are familiar with the advantages this material offers. It has a higher load bearing capacity than rubber. It is more resistant to abrasion, tearing or cutting and is unaffected by petrol, oil and ozone. It is available in a wider hardness range and can be coloured if required.

Most importantly it is much less susceptible to 'compression set' in use. When a force is applied to any flexible material it will deform but when the force is removed it should return to its original shape. After a number of deformation cycles some materials no longer return to their original shape and remain permanently deformed. This is called compression set. When this occurs in a suspension bush it will result in misalignment of the components or ovality in the bush bore, causing excess free play.

MATERIAL

Like all plastics, the material is made by linking a large number of small molecules (monomers) to produce long chain molecules (polymers). Polyurethanes use isocyanate and polyols as their monomers, although the specific substances vary depending on the required properties of the product. In fact there are two distinct types of polyurethanes.

Thermo-softening polyurethanes melt when heated so they are supplied to manufacturers in granular form to be re-melted and injection moulded. Once established, this is a very quick and cheap production process, but unfortunately this type of polyurethane is unsuitable for automotive use as it is much more susceptible to compression set and would have a very short service life if used in suspension components.

The alternative form, thermo-setting polyurethane cannot be melted once formed. In fact if heated it chemically decomposes. The polymer chains are heavily cross-linked during the polymerisation process and form a much more stable structure. Components made from thermosetting polyurethane are produced by dispensed casting, a much more expensive process, in which the monomers are accurately mixed first then a catalyst is added as the mixture is measured into the moulds. Once the mixture cures to form the required shape it can be removed from the mould although most manufacturers use an additional post-cure heat treatment to further improve the properties of the material.

Other additives can be incorporated into the mixture to prevent oxidation or hydrolysis and also to colour the material.

Different companies use different polyurethane blends to achieve the same properties but providing you buy from a

Polyurethane can by coloured as required (Polybush).

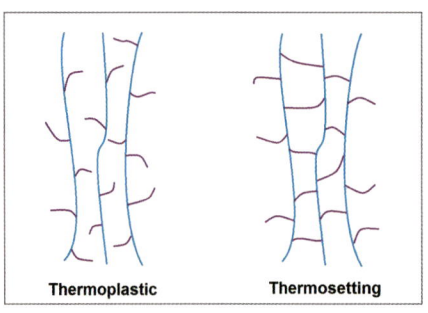

The polymer chains are cross linked in a thermosetting plastic.

Dispensed casting of thermosetting polyurethane (Powerflex).

Stainless steel crush tubes in BMW bushes (Superflex).

Rubber bushes are bonded to the inner and outer sleeves.

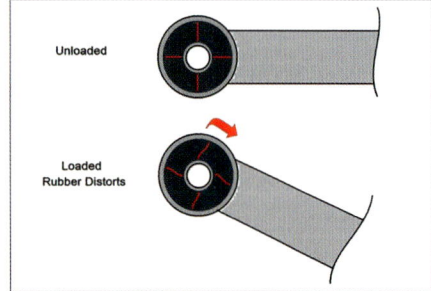

The rubber flexes to accommodate suspension movement.

Modify, Improve & Upgrade Your Kit Car

Running Gear

Polyurethane bushes are not bonded to the inner sleeve.

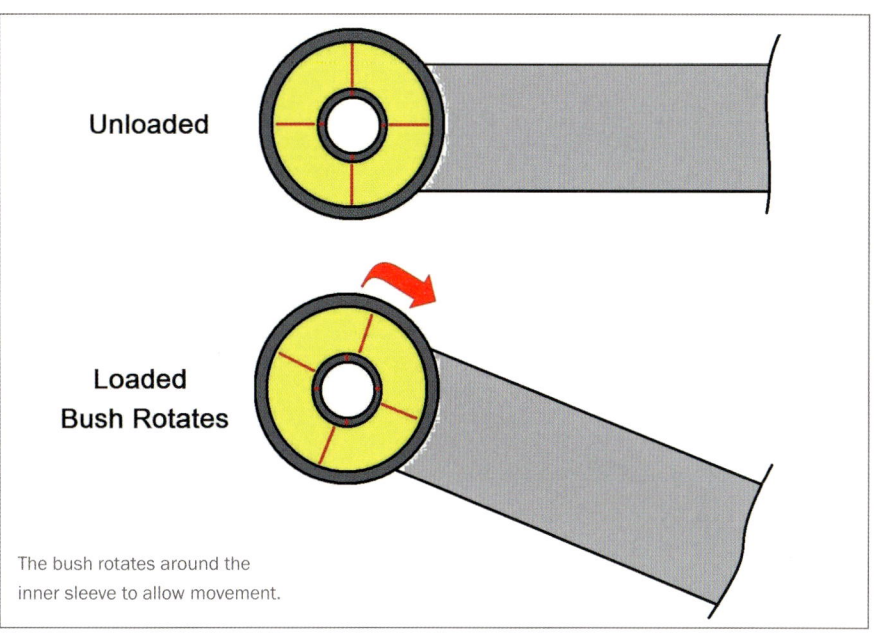
The bush rotates around the inner sleeve to allow movement.

reputable company you should be supplied with a high quality, high performance product. Beware of cheaper unbranded components though. They may be produced using lower grade materials or even injection moulded from thermoplastic polyurethane and will have a much shorter working life.

The crush tubes inside the bush can be made from zinc plated steel, stainless steel or hard anodised aluminium alloy. The polyurethane bush pivots around these tubes, so they need a smooth hard outer surface and must be corrosion resistant on both surfaces. If the outer surface corrodes it will wear the bush material badly. If the inner surface corrodes it may seize onto the clamping bolt.

HARDNESS

One of the big advantages of polymers and polyurethane in particular is the ability to tailor the material to suit its intended use. Polyurethane products for suspension use need to be produced in a wide range of hardness. The hardness of a material is measured using a Shore Hardness Gauge which basically measures the force required to push a spring-loaded needle a precise depth into the material. The Shore A scale is used for soft and semi-hard plastics and the Shore D scale is used for semi-rigid and hard plastics. There is some overlap. Shore 95A is the same as Shore 45D. Polyurethane bushes for use in vehicle suspension applications range from Shore 50A, which is soft enough to replace foam components, to Shore 95A, which is hard enough be used instead of Nylatron.

The standard OEM Cortina voided bush. It wears very quickly.

The polyurethane replacement for the voided bush (Superflex).

The spiral grooves inside the bush retain grease. (Powerflex).

A complete car set of bushes for the Healey 3000 (Polybush).

The Black Series bushes are Powerflex's competition grade.

The Comfort series are Polybush's most compliant grade.

Modify, Improve & Upgrade Your Kit Car

This MX-5 offset bush allows camber adjustment. (Superflex).

An alternative method is to use offset crush tubes. (Powerflex).

Cams are used to adjust this bush when fitted. (Powerflex).

DESIGN

Aftermarket polyurethane parts are not simply copies of the OEM components they are replacing. Not only does the material have different working properties but, when used as a suspension pivot, it operates in a completely different way. Rubber bushes are normally bonded to both their inner and outer metal sleeves so when rotation takes place as the suspension moves the rubber must flex to accommodate this.

Polyurethane bushes may or may not have an outer sleeve but either way, the bush is fixed in its housing. The inner sleeve, however, is completely free to rotate in the bush as suspension movement occurs. In this respect the bush behaves more like a plain bearing. Obviously this difference in action needs to be considered when components are being designed.

The expected motion of the bush needs to be taken into account too. Some bushes may experience only rotation, but many suspension components, such as the later Cortina rear axle voided bushes, are also designed to twist or flex in use.

Other criteria which need to be considered are the weight of the car and the expected driving style, as these will determine the forces exerted on the bush. Obviously the hardness of the bush will affect the compromise between accurate suspension location, and noise and vibration transfer, but this can also be modified by varying the relative thicknesses of the bush and the inner crush tube. A thicker crush tube inside a thinner bush will allow less unwanted movement but will produce a harder ride in exactly the same way as fitting a harder bush.

This toothed washer is rotated to adjust the bush (Powerflex).

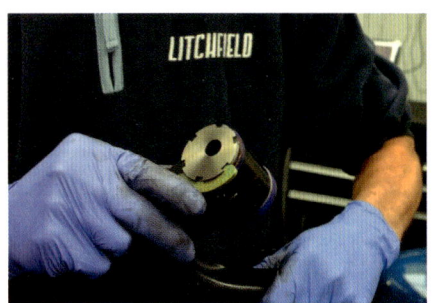

A C-spanner is used to rotate the adjuster. (Powerflex).

Polyurethane bump stops for the BMW 3-Series. (Powerflex).

Uprated BMW 3-Series trailing arm bushes. (Powerflex).

Anti-roll bar bushes for the Honda S2000 (Superflex).

Adjustable bushes for the Ford Focus (Superflex).

Polyurethane bushes for the VW Beetle spring plates (Superflex).

Running Gear

These bushes are produced specifically for the Westfield (Powerflex).

The Caterham uses a number of different bushes. (Powerflex).

This is the same bush sectioned to show the construction (Powerflex).

The Triumph trunnion bushes will fit a number of older kits. (Superflex).

This housing can be welded to custom wishbones etc (Superflex).

Bushes can be bought as a set or individual components (Superflex).

Various bump stops are available for one off applications (Superflex).

If all else fails you can fabricate bushes from polyurethane bar. (Superflex).

John made these bushes for his GTM Coupé rear suspension.

Finally, the fit of the bush needs to be considered. In order to prevent the bush rotating in its housing it has a slight (0.5mm) interference fit, but this slight compression may also affect the fit of the crush tube. There is also a slight elongation of the bush in use which is why some bushes may appear to be slightly too short when initially fitted. A more recent design innovation is the use of fine grooves cast into the inner bore of the bush or machined onto the crush tube. These grooves are designed to retain grease between the two surfaces to avoid the squeaking or creaking sounds which can occur when the bush dries out.

MATERIAL SELECTION
Most companies supplying polyurethane products will sell their products individually or as a complete car kit. The materials will have been carefully chosen by the manufacturer to produce the best performance for that particular application. A complete car kit may well incorporate a number of different hardness grades in different locations such as damper mounts, subframe mounts and suspension pivots. Depending on the intended use of the vehicle, manufacturers also offer alternative grades of material should you wish to upgrade further.

Superflex has a choice of hardness where applicable and a useful selection guide on its website to help you choose the correct grade. (www.superflex.co.uk/hardnessguide.php).

Powerflex produces its normal range for road use and its Black Series for track, rally and race use.

Polybush offers its kits in three grades known as Comfort, Dynamic and Performance, with Comfort being the softest and most compliant and the hardest Performance grade being intended for track use or for towing.

As well as incorporating firmer bushes, many of the competition orientated bush kits offer adjustable camber or caster by using offset bushes or offset crush tubes in the wishbone pivots and cam or toothed adjusters.

Although it may be tempting to assume that 'competition' bushes will make your road car handle like a racer, the general advice is to beware of going too stiff with suspension components. Hard bushes will certainly improve the location of the suspension assemblies, but they will also transmit much

Modify, Improve & Upgrade Your Kit Car **107**

more road noise and vibration into the car, giving a harsh and uncomfortable ride. Whilst this may be acceptable in a competition car on a smooth track, it can make a road car virtually undriveable on today's roads. If your car is a dual purpose road and track day vehicle, any of the suppliers will be able to advise you on the best combination for your use.

Finally it is worth considering that if suspension assemblies are taken from a heavy donor car and fitted into a lighter kit car they will automatically feel harder in use as the forces acting on them will be reduced.

UNIVERSAL PRODUCTS

Some kit cars and specials are often based around production car components and can therefore use polyurethane upgrades designed for the donor vehicle. Additionally, some suppliers have developed products designed specifically for the more common kits and specials. If neither of these options is applicable to your car, most of the suppliers offer a range of 'universal' components which can be adapted for different applications or even used to produce your own unique suspension components. These include housings, bushes, crush tubes, damper mountings and bump stops in a range of designs and sizes.

They fit in the outer end of this reversed wishbone.

If none of these are suitable, Superflex produces 12in long bars of polyurethane material in different diameters and hardness so that you can machine your own components. I did this to replace some unknown bushes on my GTM Coupé.

FITTING

Although polyurethane components are designed to be easy to fit, it is essential that the housings and locations are prepared thoroughly first. All the suppliers have comprehensive fitting guides on their websites but I have outlined the basic procedure here.

Removal of the old bush can be the most difficult part of the procedure. The central bolt may be seized in the crush tube or the outer housing may be corroded in place. For most applications the whole bush, including the outer steel sleeve, must be removed but occasionally the new bush is designed to be fitted with the existing outer sleeve still in place. Check this before you remove it.

It may be possible to push the bush from its housing using a good vice and large sockets or steel tubing. If this fails the best option is to burn out the rubber bush and inner sleeve using a blowtorch then remove the outer sleeve separately. This can be done by collapsing the sleeve inwards using a small drift or cold chisel or by carefully sawing the sleeve along its length using a hacksaw.

With the old bush removed, the housing must be thoroughly cleaned. It needs to be clear of rust and free from any dents or projections which may distort the new bush when it is pressed in.

Old bushes may push out using a vice and assorted sockets. (Powerflex).

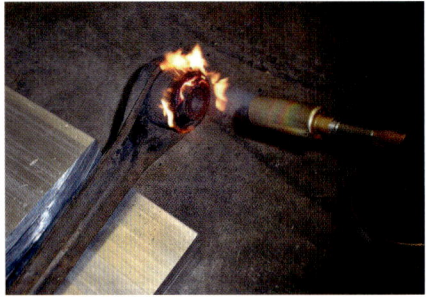

The rubber component can be burned out using a blowtorch. (Powerflex).

The inside of the housing needs to be clean and smooth.

A G-clamp or vice can be used to press in the new bushes.

Running Gear

If required grease the bush bore and crush tube (Powerflex).

Anti-roll bar bushes have been around for a while (Powerflex).

Damper eye bushes are also a common replacement. (Powerflex).

A more recent application for polyurethane is in engine mounts. (Powerflex).

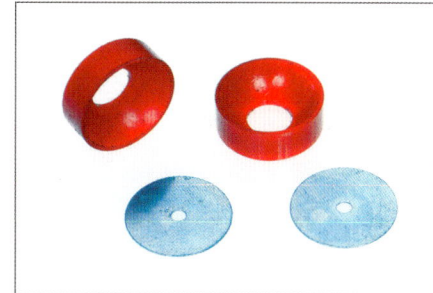
These items are for Triumph inner track rod ball joints. (Superflex).

Transmission mounts are available too. These are MX-5. (Superflex).

Polyurethane exhaust hangers will outlast rubber versions. (Powerflex).

Replacements for the easily damaged track rod end boots. (Superflex).

This bush replaces the rubber version in a Subaru gear linkage (Superflex).

The new bush is supposed to be a tight fit in the outer housing so that it cannot rotate in use. You can lubricate it with soapy water or tyre soap if necessary. These lubricants will quickly dry out leaving the bush firmly located. It may also help to warm the bush in hot water to soften it slightly. Use a vice or G-clamp to push the bush into place. Hammering the bush is not effective as the material absorbs the impact.

Once the bush is correctly located, the crush tube can be fitted. It may slide in or it may need a clamp to gently push it in. Recommendations on suitable lubricants vary, from none at all to Copper based grease, Molybdenum grease, PTFE loaded grease and Silicone grease. None of these will adversely affect the polyurethane material.

MAINTENANCE
Polyurethane bushes require no specific routine maintenance but they can be cleaned and re-greased whenever the suspension is apart for other work or if they dry out and begin to creak.

They can be cleaned with petrol, white spirit or WD-40 but alcohols (meths) and ketones (acetone or MEK) must not be used as they will degrade the material.

OTHER PRODUCTS
Most of us are familiar with polyurethane bushes, anti-roll bar supports and damper eyes but there is now a large range of polyurethane replacements for bump stops, engine mounts, steering rack components, transmission mounts, exhaust hangers, track rod end boots and gear linkages.

The only disadvantage of polyurethane aftermarket components is their higher initial cost. In reality though, if you are building a kit car, the extra cost of polyurethane over rubber is very small compared to the overall cost of the finished vehicle and the longer life and better performance also offset the price premium. If you keep the finished vehicle for any reasonable length of time it is likely that the extended service life of the polyurethane components will repay their higher initial cost.

I would like to thank Chris Wittor at Superflex for providing me with information and images used in this article and also Paul Solbe at Powerflex and Hayley Smith at Polybush for providing additional images.

Modify, Improve & Upgrade Your Kit Car

Wheel Offsets

Puzzled by deciding what wheel offset you need for your kit car?
John Dickens demystifies the measurements you need to take.

Pressed steel wheels are still found on many mass produced vehicles but alloy wheels are becoming more common even on standard family saloons. Often, this is purely for cosmetic appearance but lighter alloy wheels do offer some performance advantages.

The vast majority of kit cars are fitted with alloy wheels for the reasons described above, but there are other considerations. If you choose to fit upgraded brakes, it is possible that you will need larger diameter wheel rims in order to clear the bigger discs or calipers. If you are building a panel kit replica of a production car, you will probably find that the original wheel fitment no longer fills the wider bodywork or arches so a taller and/or wider wheel/tyre combination will be needed.

If you are considering a set of alternative wheels for your car, there are a number of dimensions which must be taken into account.

DIAMETER

This is not measured at the extreme outer edge of the rim but in the area where the tyre beading seats. If you wish to maintain the same overall diameter and rolling radius when changing wheel and tyre sizes, you also need to consider the profile of the tyres. For example, changing from 14in to 17in rims would need a switch from 185/70 by 14 to 185/50 by 17 tyres. This would keep the ride height, gearing and speedometer calibration correct. If you wish to increase the overall diameter of the wheel and tyre, this is less of a consideration.

RIM WIDTH

This is measured between the inner walls of the wheel rim. Normally when aftermarket wheels are fitted they are wider than the originals in order to allow the use of wider high performance tyres. For any given rim width, there is a range of acceptable wheel and tyre fitments so, for example, a 6in wide rim with a 14in diameter rim can accept tyre widths from 165 to 215, although the ideal fitments will be somewhere near the middle of this range of sizes. Page four of The Wheel And Tyre Bible (www.carbibles.com/tyre_bible_pg4.html) is a good source of information here.

PITCH CIRCLE DIAMETER (PCD)

This is the standard method of measuring the spacing of the wheel fixing studs or bolts. It is the diameter of a circle drawn through the centre of all the studs. There is no standard PCD. Different manufacturers use different dimensions and there may be three, four or five studs or bolts holding the wheel in place. Some popular examples....

- Ford Sierra   4 x 108mm
- Mazda MX-5   4 x 100mm
- Toyota MR2   5 x 114mm
- Classic Mini   4 x 4in
- Triumph Herald 4 x 3.75in

You must ensure that your wheels have the correct PCD for your car. It may be

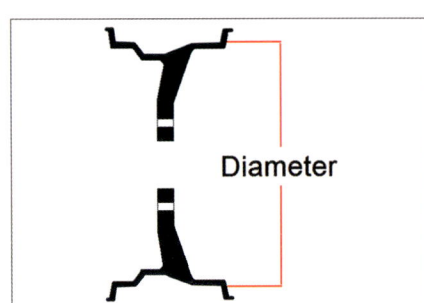

Wheel diameter is measured at the inner rim.

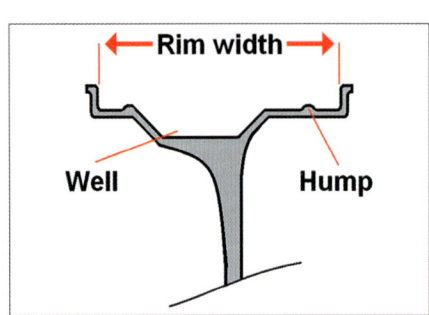

Rim width is measured between the inner rim walls.

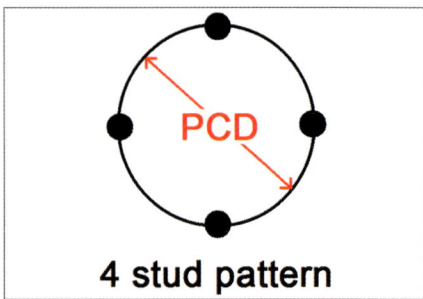

Pitch circle diameter of a 4-stud fixing wheel or hub.

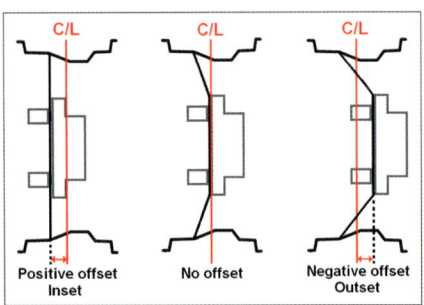

Wheel offset can be positive, zero or negative.

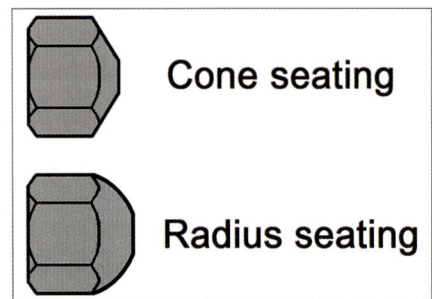

Nuts for steel wheels are shaped for wheel location.

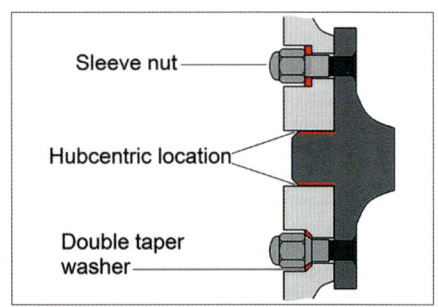

Sleeve nuts wheels an alternative locating systems.

Running Gear

Many bodykits need wheel spacers. Our project Bertini GT25 did not.

physically possible to fit 100mm PCD wheels onto a Classic Mini 4in (101.6mm) PCD hub, but they may not centre correctly and the securing studs will definitely have undesirable bending loads applied to them.

OFFSET

The centre line of the wheel rim does not normally line up with the wheel mounting face. It is normally slightly offset. The offset is described by the term ET, which is from the German word 'Einpresstiefe' translated as 'insertion depth'. If the wheel is offset away from the car's centre line, it is known as negative offset or outset, and if it is in the opposite direction it is called positive offset or inset.

WHEEL LOCATION

Pressed steel wheels are normally located by their fixing nuts or bolts. The wheels are formed with conical or radiused seats and the wheel nuts have matching contours so that as the nuts are tightened they automatically centre the wheel on the hub.

This system can also be used to locate alloy wheels, although steel washers or inserts must be used to prevent 'galling' of the soft alloy as the nuts are tightened. An alternative system for alloy wheels is to use sleeve nuts which extend through the wheel to form a positive location. These may have flat or tapered washers to protect and locate the wheel. Whichever system is used, you must ensure that you have the correct wheel nut form for your wheels with the correct thread for your hubs or studs. Modern cars tend not to use the nuts or bolts to centre the wheel. Instead they use hubcentric mounting in which a raised spigot on the hub locates in an accurately machined bore in the wheel. Once again, you must make sure that any aftermarket wheels will locate correctly on the machined spigot on your wheel hubs.

Obviously, with all these factors to consider, it is important to know exactly what you want before you go shopping for wheels. As a starting point, I suggest that you measure up the wheels you already have so that you know their overall width, diameter, PCD and ET dimensions. The first two are usually stamped or cast into the wheel, and measuring the PCD is easy enough. The ET dimension may also be marked on the rim, but if not then measuring the wheel offset is done as follows:

1. Find or cut a straight piece of metal which spans the wheel rim but not the tyre.
2. Place this straight edge across the inner wheel rim and measure from the mounting face to the straight edge. Call this dimension a.
3. Clamp or hold a flat piece of material across the rear of the centre hole in the wheel, then put your straight edge across the outer wheel rim and measure from the clamped material to the straight edge. Call this dimension b.
4. Find the centre line by adding dimensions a and b then dividing by 2. Call this dimension c.
$c = (a+b)/2$
5. To find the wheel offset (ET), subtract dimension c from dimension a.
$ET = a - c$
6. If the value is positive, the wheel has inset. If it is negative, the wheel has outset.

Once you have all the dimensions for your existing wheels, fit them back on the car and measure all the clearances around the wheel. Swing them from lock to lock to check how the clearances alter with steering movement. Now work out how much wider or taller you want your new wheels to be in order to fit the car correctly. Armed with this information you should be able to work out all the dimensions you need for your new wheels.

When making your choice, you should try to keep the ET dimension as close to standard as possible. Modern vehicles tend to use wheels with a relatively large inset as this gives stable steering under variable grip conditions (negative scrub steering geometry), but it also makes it more difficult to fit wider wheels as they can foul the suspension and bodywork at full steering lock. Decreasing the inset or increasing the

The straight edge must rest on the rim not the tyre.

Dimension a, from inner rim to wheel face.

Any flat rigid material can be used here.

Clamp material in place on the mounting face.

Dimension b, from mounting face to outer rim.

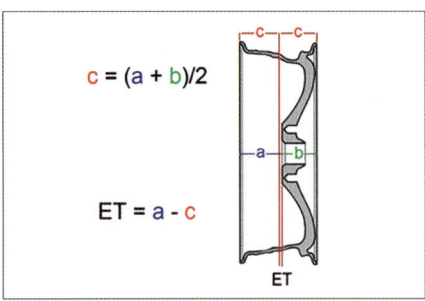
Using these dimensions you can calculate the ET.

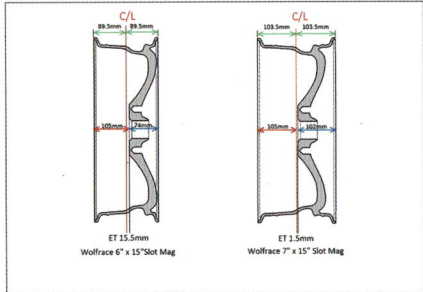
Dimensions for John's alternative Wolfrace alloys.

A thin (5mm) wheel spacer, aka a shim spacer.

Spacers are available for hubcentric wheels too.

You will need longer wheel bolts...

...if you use wheels spacers...

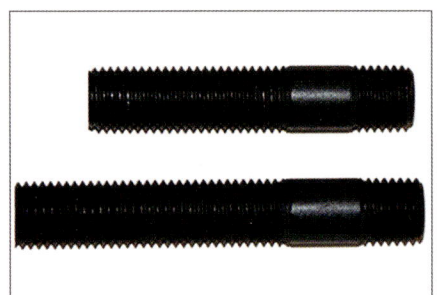
...or studs, depending on the fitment.

Bolt-on spacers can use standard studs or bolts.

Adaptors can be used to alter the PCD of the hub.

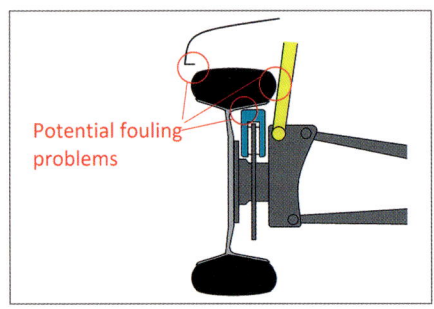
Check any new wheel/tyre combination for fouling.

Running Gear

John's Fugitive has recently had wider wheels fitted for cosmetic reasons.

outset will increase stress on wheel bearings and suspension components. At the front of the car, it can reduce steering feel, excessive kickback and instability under braking.

For purely cosmetic reasons, I have recently fitted wider wheels to the rear of my UVA and I carried out the steps described above to sort out the required fitments. Unusually, these period Wolfrace wheels accomplished the increase from 6in x 15in to 7in by 15in by adding all the extra width to the outer rim. This reduced the ET dimension from 15.5mm to 1.5mm. This is acceptable on a rear wheel, but if fitted to the front of the car the small offset could well produce a steering system with very little feel or feedback.

Manufacturers do not produce an infinite number of wheel sizes. It may be that the diameter or width you need is simply not available with the PCD or offset that your car requires. If this is the case, all is not lost. Wheel spacers can be used to increase the ET dimension if the available wheels have too much inset. Spacers are available from around 5mm thick to 30mm thick, but personally I would regard 25mm as the safe limit. They can be plain or hubcentric. If you choose to fit spacers, you will also need longer wheel bolts or studs too. Remember, though, that fitting any spacer will reduce the inset or increase the outset by the thickness of the spacer, so be careful not to stray too far away from the manufacturer's figure. A third type of spacer is the bolt-on. This is fitted using the original studs or bolts and has a second set of fasteners to attach the wheel. I have never used these as I do not really trust them, but many people are quite happy with them. If you are really having trouble finding suitable wheels, it is possible to change the PCD of your car by using adaptors. These are bolt-on spacers which have different PCD dimensions for the two sets of fasteners. Using adaptors, it is even possible to convert from four to five-stud mounting or vice versa. Unfortunately, in order to accommodate the mounting nuts or bolts, these adaptors tend to be available in only the thicker versions.

When you have finally purchased your new wheels and fittings, mount them on the car, without tyres, and check for any fouling on the brakes, suspension or bodywork over the full range of steering and suspension movement. If there is a problem at this point it may still be possible to exchange them. If all is well, get the tyres mounted and do the same checks all over again.

Finally, remember that any change in overall diameter of the wheels will mean that your speedometer will no longer read correctly and will need recalibrating. Depending on the type of instrument you are using, this may be just a case of inserting one value in place of another in a digital system, or the use of a professional service to recalibrate an analogue instrument.

CVs & Driveshafts

John Dickens shows you how to sort out old driveshafts and replace ageing CV joints.

Driveshafts need flexible couplings to accommodate suspension travel.

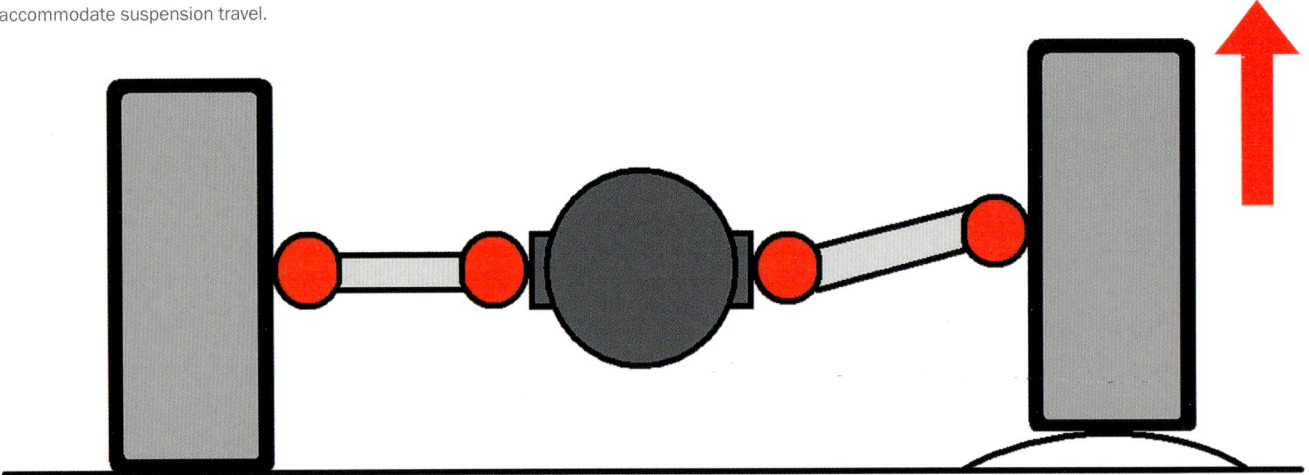

Cars with independent suspension of the driven wheels normally use external driveshafts to carry the power from a central chassis mounted differential unit to the wheels. These driveshafts need flexible couplings at each end to accommodate the movement of the suspension and, if the vehicle is front-wheel drive, steering movement too. Unless the vehicle has swing axle suspension, where the driveshaft is also used as a locating link, the shaft also has to accommodate some change in length or 'plunge' as the suspension moves.

The most efficient type of flexible coupling for this application is the constant velocity or CV joint. It resembles a large ball bearing with an inner race splined to the driveshaft, a number of large ball bearings located by a cage and an outer race which is connected to the stub axle. The ball bearings transmit the power from the inner to the outer race whilst allowing some angular and linear movement.

CV joints have a very long service life providing they are kept correctly lubricated. They are normally sealed by a flexible rubber boot and the most common cause of failure is a split or perished boot allowing grease to leak out and water to leak in. An early sign of failure is a clicking or knocking noise from the joint. At this stage the joint may be salvageable by cleaning and re-greasing but only dismantling and cleaning for thorough inspection will confirm whether the joint can be reconditioned or must be replaced. The procedure outlined here, using the Beetle CV joints from my UVA, shows how this is done.

My car has two identical driveshafts with four identical CV joints. At some stage a previous owner has replaced the original boots with some rather bright aftermarket polyurethane alternatives. Hopefully, if he

Beetle driveshafts from John's UVA. Boots not OEM!

These particular CV joints are very shallow.

Joint still packed with correct olybdenum grease.

114 *Modify, Improve & Upgrade Your Kit Car*

Running Gear

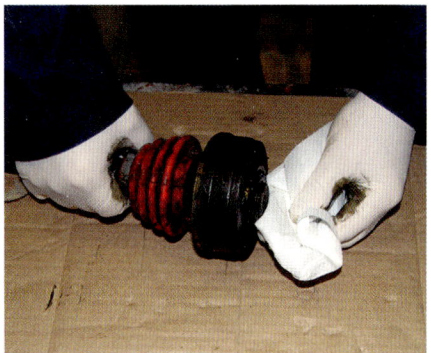
Wipe away as much grease as possible before the strip down starts.

This circlip retains the joint on the shaft. Other systems are used too.

The Mini CV joint is retained using an internal snap-ring.

Mark the parts so that they can be refitted in the same alignment.

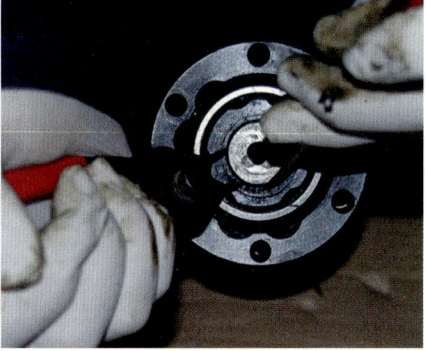
This type of circlip needs to be wound out of its groove.

Take care not to damage the circlip if it needs to be reused.

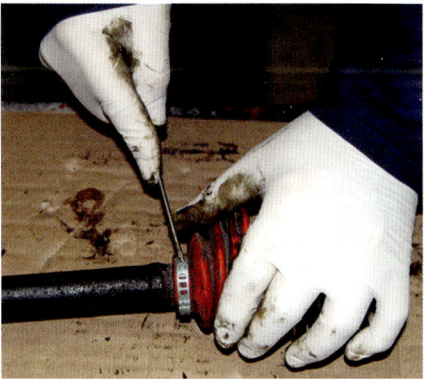
This clip can be released. Some need to be cut.

Slide the boot out of the way to access rear of joint.

This particular joint needs to be drifted off shaft.

Dished washer must be fitted correct way round.

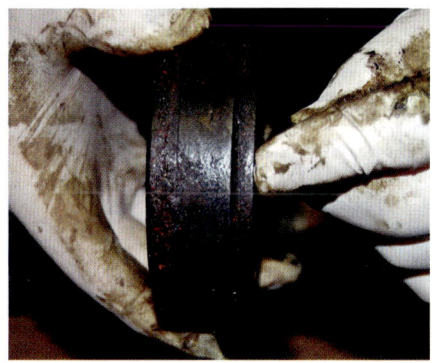
This stepped edge faces the end of the shaft.

This chamfer faces towards the centre of the shaft.

Modify, Improve & Upgrade Your Kit Car **115**

The non-chamfered face fits against the circlip.

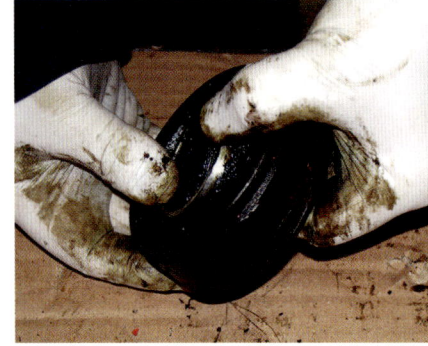
Inner race and cage can be tilted in the outer race.

Ball bearings are accessible when the cage is tilted.

The ball bearings come out easily.

Inner race and cage can be removed from the joint.

Fully disassembled CV joint ready for cleaning.

The old boot is removed and discarded. Hurray!

Shaft and splines cleaned up for reassembly.

About 250ml of Jizer cleaned all four CV joints.

re-greased the joints at the same time they should still be in reasonable condition but, since I am doing a full rebuild on the vehicle, I will strip and check each joint to be sure.

If you are servicing more than one CV joint at the same time take great care not to mix the components. They may be manufactured as identical units at the factory but they wear differently in use and interchanging components may cause early failure.

The VW CV joints are quite shallow in construction and bolt to the drive flanges using high tensile 12.9 grade bolts. The joint shown here is still well greased. The dust and strands of fibreglass are the result of the joint being left open in my garage for some time.

STRIP DOWN

The first step is to wipe away as much of the grease as possible so that the mechanism can be seen clearly. This joint is retained by an external circlip but some joints, like the Mini CV joint, are retained by a circular wire 'snap-ring' inside the joint. If this is the case the joint can usually be removed by striking it firmly with a hide or copper faced hammer. This normally compresses the snap-ring enough to free the joint.

Before removing the circlip I am marking the individual components with a scriber so that I can reassemble them in the same alignment. Using circlip pliers and a small screwdriver, the circlip is prised out of its groove and removed from the shaft. Once the retaining clips are prised open or cut the rubber boot can be slid out of the way. At this stage some joints may simply slide off the driveshaft but the VW splines are an interference fit in the joint so it needs to be drifted off. The best method is to support the inner race in a vice and drift out the shaft but it can also be done by gripping the shaft and carefully drifting against the inner race.

When the joint is free from the shaft check for and remove any remaining components. This particular assembly has a concave washer which acts as a spring and shock absorber. It must be refitted the correct way round.

116 *Modify, Improve & Upgrade Your Kit Car*

Running Gear

These are the joint components after cleaning.

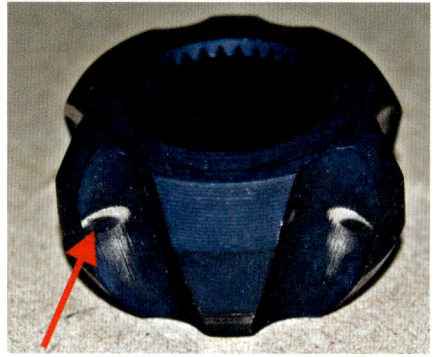

A groove has been worn in each track of inner race.

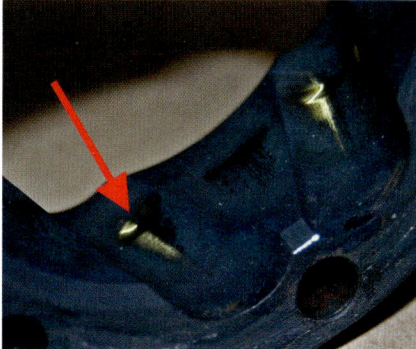

Outer race has suffered wear too. Joint is notchy.

Before reassembly make sure your alignment marks are visible.

You need to wiggle the inner race into the cage to fit it.

Line up the two components then fit them to the outer race.

Make sure your alignment marks are correct.

Tilt cage and inner race so ball can be pressed in.

Repeat this for the rest of the ball bearings

Whilst it is possible to clean, inspect and re-grease the joint while it is still assembled it is much easier to do this with the joint fully stripped. Before stripping the joint check how the components are aligned. On this joint the step on the outer race fits into the drive flange on the stub axle or transaxle. The chamfered edge on the inner spline fits against the concave washer and the non-chamfered edge fits against the circlip. Make notes or take photos of your CV joint so that you can re-assemble it correctly.

To strip the joint tilt the inner race and cage until the ball bearings are exposed.

They may fall out easily or they may need to be gently prised out with a small screwdriver. Once a few of the balls have been removed the inner race and cage can be withdrawn from the outer race and the remaining balls removed. When all the ball bearings have been removed from the cage it can be wiggled off the inner race, leaving the joint fully disassembled and ready for cleaning. It is possible to re-use an undamaged rubber boot but, considering the price of a replacement, it is false economy to do so. Slide the old boot off the shaft and discard it. The exposed drive shaft and splines can now be cleaned up thoroughly.

CLEANING/INSPECTION

All the old grease needs to be removed from the joint and shaft. Most people will have their own preferred method for this. I have tried petrol and similar solvents, water based cleaners and water washable degreasers and I always come back to Jizer by Deb. It dissolves oil and grease like a petroleum based solvent but washes away with water. Along with a selection of old paintbrushes and toothbrushes, this will tackle most jobs.

A slight greasy film sometimes remains after rinsing. This can be removed using a solution of washing powder in water if you wish to paint the cleaned item.

This shows how far the inner race can move out...

...and in.

Take care not to damage rubber boot on splines.

Push boot well out of the way for the time being.

This assembly needs this concave washer.

Carefully line the joint up with the drive splines.

Carefully tap inner race to drive joint onto shaft.

Socket or piece of tube is needed to fully seat joint.

Driveshaft should protrude so circlip groove is seen.

Once all the components are clean they can be inspected for wear or damage. Check the ball bearings, inner races and outer races for signs of pitting, grooves or ridges. The balls should slide easily in the tracks without any binding. If any wear is evident renew the joint.

This is not a job you would want to do frequently so fit a new joint now and it will probably last the life of the car. Currently a Mini CV joint is around £30, a VW Beetle joint is £40 to £50 and a Ford tripod joint is about £40. As can be seen from the photos, there is wear evident on both the inner and outer races of my CV joint so in spite of my earlier optimism it will need to be replaced.

RE-ASSEMBLY

If your CV joint is in good condition it can be re-assembled, greased and put back into service. First identify the alignment marks that you made when you stripped the joint. Begin the assembly by wiggling the inner race into the cage. If it seems reluctant to engage try turning the cage round so that the race is entering from the other side.

On my joint the cage had a 1mm larger aperture on one side and assembly was only possible through this larger opening. Line the cage and inner race up then place the assembly into the outer race making sure that all your alignment marks line up. Tilt the inner race and cage then push one ball into position and realign the three components. Tilt the cage and race again and push in the second ball. Continue with this procedure until all the balls are located correctly in the joint. You may need to use a little more effort as the joint fills up. Check that the fully assembled joint moves freely. This particular joint also allows a few millimetres of linear movement as shown in Figures 35 and 36.

You could grease the joint at this stage but it would make the rest of the assembly very messy. In this case I decided to fit the joint to the driveshaft first. Don't forget to fit the boot and any other components onto the shaft before you fit the joint. Taking care not to damage the rubber boot on

118 *Modify, Improve & Upgrade Your Kit Car*

Running Gear

This type of circlip is wound onto the shaft.

When positioned, circlip locates fully in groove.

Use plenty of newspaper before you open grease.

Sachet of grease often provided with a new joint.

Molybdenum grease can also be bought separately.

Add the grease in golf ball sized lumps.

Work the grease fully into one side of the joint.

If accessible, repeat process at rear of the joint.

Clean away excess grease and fit protective boot.

the splines ease it onto the driveshaft and push it well down the shaft out of the way. A little grease will help it slide into place.

This assembly has a concave washer under the joint so this was fitted next. Making sure the CV joint is the correct way round line it up on the splines and push it into place. It may slide easily into place or it may need to be drifted on. Drift on the inner race only or you could damage the bearing surfaces. I needed to use a socket for the last few millimetres as my driveshaft protrudes through the joint. The final step in assembling the driveshaft is to re-fit the circlip. With the new or reconditioned CV joint fitted to the drive shaft the final step is to pack the joint with grease.

The correct grease for CV joints is lithium based but has a molybdenum disulphide additive. As a result the grease is dark grey making the whole process seem much messier than, for example, wheel bearing work. Before you begin spread plenty of newspaper over the working surface (Fig 46) and make sure you are wearing rubber gloves. Normally when you buy a replacement CV joint a small sachet of the correct grease is included.

If you are very lucky you may even get this grease provided with a simple boot kit too. As I was fully reconditioning four joints I bought a 500g tub of CV joint grease from Halfords. Start with a blob of grease about the size of a walnut and work it into the joint. Continue adding more grease, tilting the joint regularly, until one side of the joint is completely packed. Repeat the process working from the rear of the joint if accessible. When you are happy that the CV joint is fully packed with grease clean away the excess and fit the rubber boot. The driveshaft and joint is now ready to be re-fitted to the car.

This procedure, although messy, is not particularly difficult but it is very important to make notes and take photographs as you strip the driveshaft and joint so that you can get all the components in the correct order and alignment when you re-assemble them.

Modify, Improve & Upgrade Your Kit Car

Rollover Protection

Some form of rollover protection may one day save your life. **John Dickens** explains your options.

A single braced roll hoop for the driver.

A plain un-braced roll bar.

A basic roll cage in this 1970 Corvette.

The roll hoops are un-braced in this Barchetta 595.

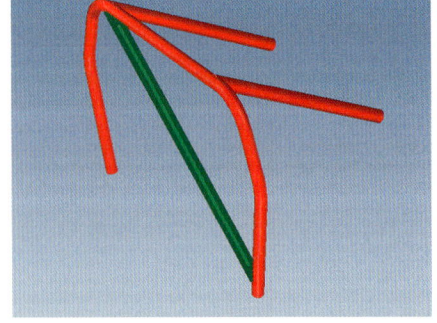
Main roll bar with rear bracing and a diagonal strut.

This has front and rear bars with linking roof bars.

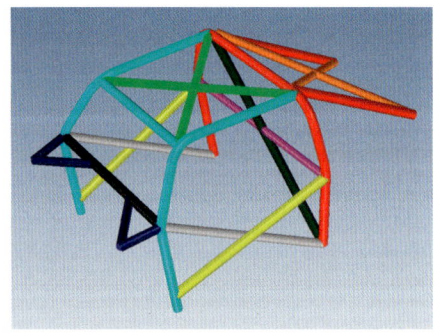
A full cage with 'X' braces and dash bars.

Rollcages

This cage is certified to FIA and MSA specification.

A CMM in use to design a roll cage.

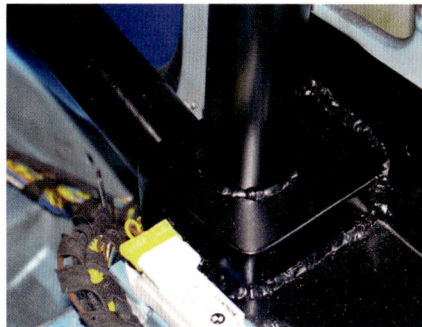
Bars and cages can be welded into car permanently.

Alternatively the roll structures may be bolted in.

The CDS and T45 tubing stock at Custom Cages.

A CNC bending machine.

As one of the more senior members of the CKC team, I can remember back to a time when there was a relatively relaxed attitude to car passenger safety. Seatbelts became mandatory in new cars in 1968 but it was not until 1983 that it became compulsory for front seat occupants to wear them. It was as late as 1997 that, despite opposition from the manufacturers, Euro NCAP results were first published with the Volvo S40 achieving the first 4-star rating.

Once manufacturers realised that customers were taking note of the NCAP rating their attitude quickly changed and safety became a selling point and therefore a priority.

Kit cars are not tested by NCAP and, thankfully, in the UK there is no requirement for the crash testing and type approval which effectively killed off kit cars in most other European countries. Instead we have the IVA test which determines the safety of individual cars by examining, among other things, their general engineering, their seats and mounting points and, in particular, the mountings for the seat belts.

Interestingly neither NCAP nor IVA testing prioritises occupant protection in the event of a car rolling over although, to their credit, the passenger safety cells produced by most mainstream manufacturers fare relatively well in the event of their cars overturning due to the very effective roll structure formed by the strong screen pillars front and rear. Convertibles are more problematic in this area though. Mass produced convertibles tend to have very strong windscreen structures designed to act as roll bars (a feature also used by the early Marlin kits) but may or may not have additional roll protection behind the seating area. The Mazda MX-5, for example, has two small roll hoops but the Toyota MR2 and the MGF do not.

Before going into more detail about roll over protection it might be a good idea to outline the terms I will use to describe the various components as some of the terms are often interchanged.

Roll hoop – This is essentially a narrow roll bar spanning the width of a single occupant's shoulders. They are often fitted in pairs.

Roll bar – This is a single bar that extends across the width of the car protecting both driver and passenger.

Roll cage – This is a specially constructed frame built inside the passenger compartment. It usually consists of front and rear roll bars linked by roof and door bars and may also include diagonal or X bracing between the structural members.

As far as road going kit cars are concerned, the choice of whether or not to fit roll over protection is entirely up to you. There is no compulsion to fit a roll bar or even a windscreen to a road car. In this more safety conscious era though, very few people choose to go without this safety feature and most owners choose to fit at the very least a simple roll bar or a pair of roll hoops.

Most kit car manufacturers offer one or more forms of roll protection depending on the intended use for the car, but make sure you know exactly what you are buying as some of these structures are intended to be cosmetic rather than functional.

If you intend to compete in any form of motorsport then your choice is much more restricted. Some form of roll protection is likely to be compulsory, no matter what class you compete in, and the details of exactly what is required will be given in the regulations governing the particular competitions you wish to enter. In the UK the competition organisers will normally refer competitors to the roll bar regulations found in the MSA Blue Book which is effectively the 'bible' for all competitors. It contains diagrams and dimensions for all forms of roll protection along with material specifications, drawings of mounting systems, jointing methods and clearances.

As a minimum you will probably be required to fit a full roll bar, front or rear bracing and a diagonal support. Some

Modify, Improve & Upgrade Your Kit Car **121**

classes may require this to be augmented with a front roll bar around the screen and roof bars joining the two main bars to form a cage.

If you have seen racing Caterhams you may have noticed that they also use external lower linking bars which give some side impact protection too.

Rally cars normally require extensive roll structures which also tie together suspension mounting points and effectively form an additional spaceframe chassis.

PROFESSIONAL DESIGNS

Any form of roll protection used in motorsport needs to be certified as being to the required standard. The two standards in operation in the UK are from the FIA or the MSA. They are very similar but differ in detail. Most companies produce cages which meet both standards. The requirement for certification effectively means that, for competition use, you need to have your roll cage professionally made and fitted, although some companies will certify their cages after home assembly subject to certain checks. Custom Cages, for example, sends out a weld check sample with its kits. The builder must weld up the sample then return it for testing along with photographs of the fitted cage. If the welds are satisfactory and the cage has been fitted correctly a certificate of conformity will be issued.

Most of the companies manufacturing roll protection can design and build MSA/FIA certificated cages as bespoke items for individual cars too, so kit cars will cause them no problem.

I contacted Alec Bedford at Custom Cages and Andy Robinson of Andy Robinson Race Cars for more information about their bespoke services. Both these companies are very experienced in producing one-off cages for motorsport, classic, and custom vehicles and offer a range of services. Not surprisingly, both companies adopt similar procedures.

The first step is to inspect the car for suitable mounting points. It may be that additional plates need to be welded onto the chassis. It may even be necessary to add additional bracing to the chassis to cope with the extra loads which may be imposed. For composite cars without a separate chassis it will be necessary to fabricate load spreading plates which sandwich the composite panels. Andy had to adopt this approach when faced with an early Marcos Coupé with its plywood torsion box chassis.

The next step is to accurately measure the interior of the car and the chosen mounting points. Alec uses a co-ordinate measuring machine (CMM) to produce an accurate 3D representation of the car interior and has recently invested in 3D laser scanning

Roll benders can produce long smooth bends.

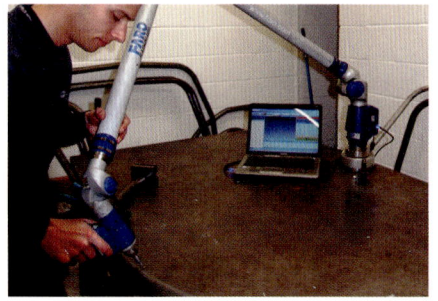
Checking the accuracy of a roll bend using a CMM.

This plate feeds loads into the floor and sill.

The notched tubes are tack welded first.

All joints are then fully welded.

TIG welding: strong welds with minimum distortion.

Set of cut, notched and bent tubes ready to assemble.

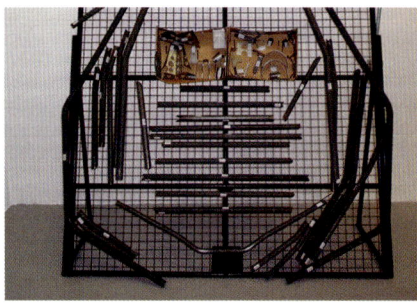
Tubes can be fitted in-house assembled by client.

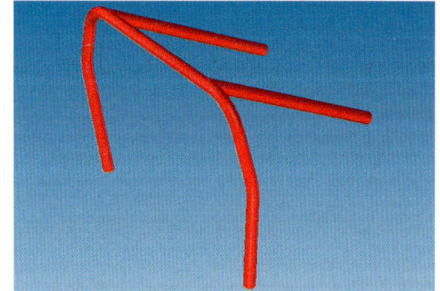
Minimum spec John would consider in his cars.

Rollcages

equipment for even greater accuracy. The measurements are then used to generate a 3D CAD image of the car interior. Andy pointed out that at this stage, before the design process commences, other factors need to be taken into account.

The intended use of the car is important as, for road use; some sections may need to be removable or left out altogether. The cage may be welded in permanently or it may need to be bolted in place. The physical size of the occupants must be considered too. Once the design of the cage is finalised, the material is chosen. Both companies work in cold drawn seamless steel (CDS) and T45 chrome manganese steel (lighter for the same strength). Andy also uses 4130 chrome molybdenum steel as this is required for the drag racing cages he produces.

The tubes are then cut to length, notched as required and bent on numerically controlled bending machines. Roll benders are used to produce long smooth bends which can follow the contours of a body shell exactly. Mounting plates are cut and bent so that they feed the loads into structural areas of the car or chassis.

The final step is to assemble the cage in the car. Both companies use TIG welding to produce strong consistent welds with the minimum of distortion.

Depending on the complexity of the roll bar or cage the car will be required for seven to 14 days to complete the whole process.

John built his own roll bar for his Spartan racer (a while ago).

Alternatively, if the customer can supply the dimensions, both these companies can produce bars or cages in kit form with the tubing bent, cut to length and notched for the customer to assemble.

The dimensions can be supplied as 3D CAD data, drawings or templates. The cost of a bespoke bar or cage will vary depending on the design and the materials chosen. T45 tubing is around twice the price of CDS steel so this will add to the cost of the finished article.

DESIGNING YOUR OWN BARS

If your car will be for road or track day use only, there is no necessity for any form of certification so you are free to design and build your own system. It makes sense though to use the RAC Blue Book designs as a guide if you hope to produce an effective roll bar.

I would suggest that a minimum would be a roll hoop and two rearward braces. You also have a free choice of materials but once again the best advice would be to stick with custom and practice. I have actually seen roll bars made from nothing more than 18swg exhaust tubing which would obviously be totally useless in a crash.

When I raced in the 750 Motor Club Kit Car Challenge series many years ago, certification was not required and I made my own roll bars for both my Spartan and my Mini Marcos racers. I used 50mm outside diameter Blue Band industrial plumbing pipe which has a wall thickness of 3mm and was more than strong enough for my needs. I had the bends done at the suppliers where I bought the pipe and I welded up the assembly myself. There is nothing to stop you choosing this route too but, depending on your abilities, you may prefer to draw out your proposed design and have it cut, and bent by a professional company for you to weld up and fit yourself.

The shape of the roll bar is also up to you, but in order to provide adequate protection it must obviously be higher than your

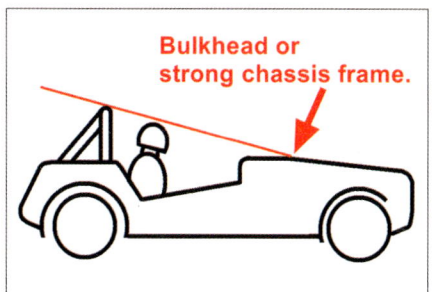
Driver's head mustn't touch road with car inverted.

Weld-in cage follows interior contours very closely.

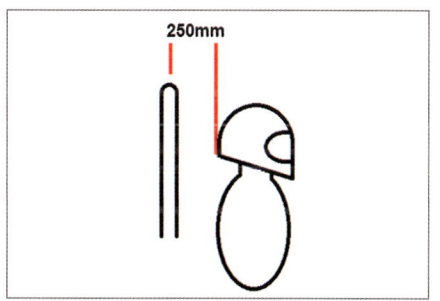
Minimum clearance between head and roll bar.

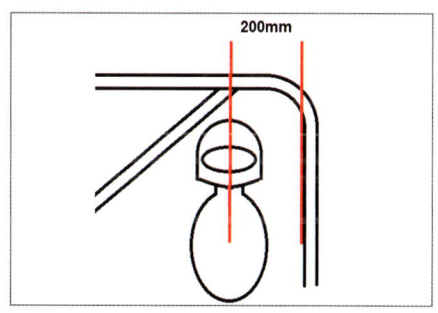
Minimum clearance for head and shoulders.

Modify, Improve & Upgrade Your Kit Car **123**

Bolt-in cage needs clearance for fit and removal.

Simple and commonly used bolt-in joint design.

Joint is used here to make a door bar removable.

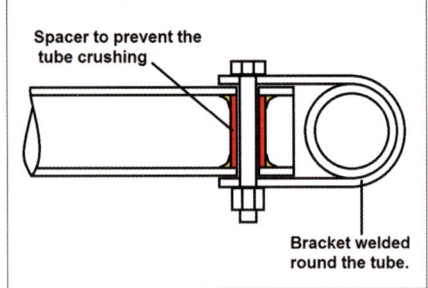
A simple wrap-round bracket can be used.

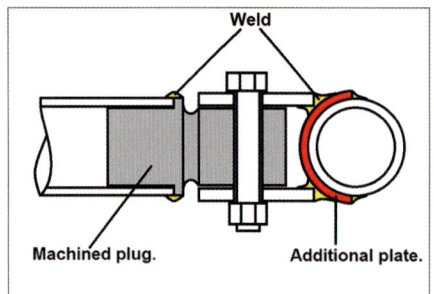
This is a more complex machined joint.

These roof bars use another type of bolt-in joint.

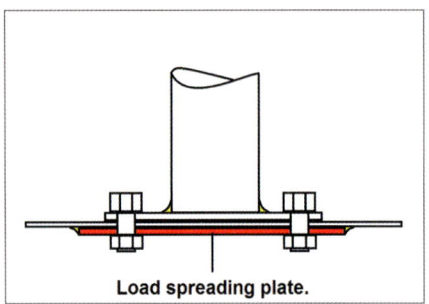
Thin sections need to be reinforced by load spreading plates.

Composite structures need large load spreading plates.

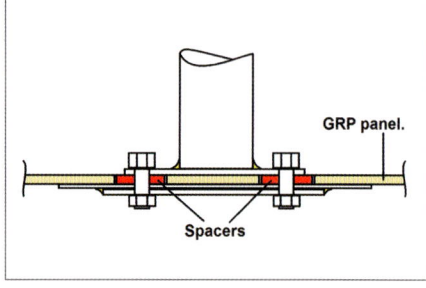
Spacers through uneven GRP to ensure metal to metal contact.

head and wider than your shoulders when normally seated. It should also prevent you touching the ground should the car end up inverted. Once again the Blue Book has a series of diagrams which should give you the idea.

Professionals will normally design the roll bar or cage so that it follows the contours of the car interior as closely as possible. This allows for the minimum intrusion into the interior space and makes the installation as unobtrusive as possible. Remember to allow for the thickness of any padding that you intend to fit too.

MAKING YOUR OWN BARS

If you decide to bend up your own bar, bear in mind that it can be difficult to achieve exact dimensions when bending large radius tubing. But fortunately software is now available to enable you to calculate the bending allowances you need to use. There is also software which will allow you to print out a template to accurately notch the ends of supporting tubes prior to welding them in place.

TIG welding is ideal but MIG welding is perfectly acceptable too. Gas and arc welding can create too much distortion in the welds so they are best avoided if possible. If you use your car both on the road and for track days you may want to make some of the structure, such as the diagonal brace or door bars, removable by using one of the bolt-in joint designs.

Your roll cage can be welded or bolted into the car. If bolts are used the MSA specification calls for a minimum of three M8, 8.8 grade bolts at each mounting point, but most professionals use four. The mounting points themselves should be as strong as possible. Ideally large mounting plates should be welded to the chassis during the build. Thin steel panels should have load spreading plates on one or both sides of the mounting point and fibreglass shells should have even larger plates bonded in or sandwiching reinforced composite structures. If the GRP body is mounted to a steel chassis spacers could be fitted through the GRP so that metal to metal contact is achieved in the mount.

If you are designing and making your own roll bar always work on a worst case scenario. It may seem pessimistic but always bear in mind that you need to produce a structure which is strong enough to prevent your head from contacting the road surface whilst the car is sliding upside down at high speed. Have a look at a few examples to see what is commonly used and, if in doubt, use the Blue Book as your guide.

Appendix: Useful Contacts

This list includes many suppliers you'll find useful while modifying, improving and upgrading your kit car.

BRAKE PARTS (DISCS/CALIPERS/PADS)
AP Racing T: 024 7663 9595.
W: www.apracing.com
EBC Brakes
W: www.ebcbrakes.com
Ferodo W: www.ferodo.co.uk
HiSpec Motorsport
T: 01322 286850.
W: www.hispecbrake.co.uk
Mintex W: www.mintex.co.uk
MNR T: 01423 780196.
W: www.mnrltd.co.uk
Pagid W: www.pagid.com
Wilwood (via Rally Design)
T: 01227 792792.
W: www.rallydesign.co.uk

CARBURETTOR/INJECTION SERVICES/ENGINE MANAGEMENT
DanST Engineering T: 07921 168507.
W: www.danstengineering.co.uk
Omex Technology T: 01242 26065.
W: www.omextechnology.co.uk
Race Technology T: 01773 537620.
W: www.race-technology.com
Webcon T: 01932 787100.
W: www.webcon.co.uk

COOLING SUPPLIERS (RADIATORS/HOSES)
Forge Motorsport T: 01452 380999.
W: www.forgemotorsport.co.uk
Pacet T: 01628 526754.
W: www.pacet.co.uk
Pro Alloy Motorsport T:0845 226 7561.
W: www.proalloy.co.uk
Radicool Fabrications T: 01280 701350.
W: www.radicool-fabrications.co.uk
Samco Sport T: 01443 238464.
W: www.samcosport.com
Silicon Hoses T: 0845 838 5364.
W: www.siliconhoses.com
Viper Performance T: 0845 0953 423.
W: www.viper-performance.co.uk

ELECTRICAL/WIRING SUPPLIES
Auto Electric Supplies
T: 01584 819552.
W: www.autoelectricsupplies.co.uk
Autocar Electrical T: 020 7403 4334.
W: www.autocar-electrical.com
Autosparks T: 01423 506133
IEM Services T: 01209 214086.
W: www.thewiringproject.co.uk
Vehicle Wiring Products T: 0115 9305454.
W: www.vehicleproducts.co.uk
World of Wiring T: 01782 208050.
W: www.blitzworld.co.uk

ENGINE SPECIALISTS/ENGINE PARTS
AB Performance T: 01449 736633.
W: www.abperformance.co.uk
Avonbar T: 01279 873428.
W: www.avonbar.com
Burton Power T: 020 8518 9189.
W: www.burtonpower.com
Cambridge Motorsports Parts
T: 01462 684300.
W: www.cambridgemotorsport.com
Cat Cams T: 01444 243720.
W: www.catcams.co.uk
Dee Ltd T: 01926 311915.
W: www.dee-ltd.co.uk
Dunnell Engines T: 01449 677726.
W: www.dunnellengines.com
Holeshot Racing T: 028 3882 0026.
W: www.holeshotracing.co.uk
ITG T: 024 7630 5386.
W: www.itgairfilters.com
LS Power T: 01949 843299.
W: www.gdcars.com
Partsworld Performance
W: www.partsworldperformance.com
Performance Unlimited
T: 01904 489332.
W: www.performanceunlimited.co.uk
Piper Cams T: 01303 245300.
W: www.pipercams.co.uk
QEP (Cat Cams) T: 01444 243720.
W: www.q-e-p.co.uk
Real Steel T: 01895 440505.
W: www.realsteel.co.uk
TTS Performance T: 01327 858212.
W: www.tts-performance.co.uk
Ultimate Performance T: 01604 771221.
W: www.ultimatep.com
Yorkshire Engine Supplies
T: 07960 011585.
W: www.yorkshireengines.co.uk

ENGINE MANAGEMENT
Autocar Electrical Equipment (Lumenition)
T: 020 7403 4334.
W: www.lumenition.com
KMS T: +31 (0) 402854064.
W: www.van-kronenburg.nl
Omex Technology T: 01242 260656.
W: www.omextechnology.co.uk
Trigger Wheels E: sales@trigger-wheels.com
W: www.trigger-wheels.com

EXHAUST PARTS/FABRICATION
Custom Chrome T: 024 7638 7808.
W: www.custom-chrome.co.uk
Simpson Race Exhausts
T: 01753 532222.
W: www.simpsonraceexhausts.com

FIBREGLASS REPAIRS
CFS T: 01209 821028
W: www.cfsnet.co.uk
Dynamic Mouldings
T: 01454 222 899.
W: www.dynamicmouldings.co.uk
East Coast Fibreglass
T: 0191 497 5134.
W: www.ecfibreglass.co.uk
GW-GRP Designs T: 01507 524426.
W: www.gw-grpdesigns.co.uk
Westgate Composites
T: 07733 282947.
W: www.westgatecomposites.com

FUEL/OIL/BRAKE FLUID COMPONENTS
BGC T: 01945 466690.
W: www.bgcmotorsport.co.uk
Earls T: 01803 869850.
W: www.earls.co.uk
Hosetechnik T: 0845 838 5364.
W: www.hosetechnik.com

GEARBOX SPECIALISTS
3J Driveline T: 01926 650426.
W: www.3jdriveline.com
BGH Geartech T: 01580 714114.
W: www.bghgeartech.co.uk
CG Motorsport T: 01132 426359.
W: www.clutch-specialists.co.uk
Elite Racing Transmissions
T: 07976 487861.
W: www.eliteracingtransmissions.com
MST Developments T: 07890 587531
Quaife T: 01732 741144.
W: www.quaife.co.uk
Tran-X T: 01732 741144.
W: www.tran-x.com

Modify, Improve & Upgrade Your Kit Car

GEARBOX SPECIALISTS (REVERSE)
Elite Racing Transmissions
T: 07976 487861.
W: www.eliteracingtransmissions.com
Lynx AE T: 01908 510000.
W: www.lynxae.co.uk
MNR reverse box T: 01423 780196.
W: www.mnrltd.co.uk
Quaife T: 01732 741144.
W: www.quaife.co.uk
Westgarage Engineering T: 01383 850480.
W: www.westgarage.co.uk

INSTRUMENT/GAUGE SUPPLIERS
Acewell T: 0191 640 8663.
W: www.acewell.co.uk
Digital Speedos T: 07967 676703.
W: www.digitalspeedos.co.uk
ETB Instruments T: 01702 601055.
W: www.etbinstruments.com
Race Technology T: 01773 537620.
W: www.race-technology.com
Racetech W: www.racetechdesign.com
Revotec T: 01491 824424.
W: www.revotec.com
Smiths (via Europa)
T: 01283 815609.
W: www.europaspares.com
SPA T: 01827 300150.
W: www.spa-uk.co.uk
Stack W: www.stackltd.com
Trailtech T: 01896 753111.
W: www.trailtech.net

KIT CAR BUILDING SERVICES
Arden Automotive T: 01235 813161.
W: www.ardenautomotive.co.uk
Automotive Solutions and Racing
T: 01773 719287.
W: www.kitcar.me.uk
Birch Brothers T: 01274 834921.
W: birchbros.org.uk
Thunder Road Cars T: 020 8502 4090.
W: www.thunderroadcars.com
Southways Automotive T: 07976 531824.
W: www.southwaysautomotive.co.uk
Sussex Kit Cars T: 01435 812706.
E: john@sussexkitcars.co.uk

LIGHTING
SVC T: 0845 658 1251. W: www.s-v-c.co.uk

MISCELLANEOUS
Aluminium fabrication – Bogg Brothers
T: 01944 738234. W: www.boggbros.co.uk
Aluminium fabrication – Alloy Racing Fabrications T: 01623 835805.
W: www.alloyracingfabrications.com
Carbon Mods T: 01782 324000.
W: www.carbonmods.co.uk
Heater – T7 Design T: 07595 975777.
W: www.t7design.co.uk
Powdercoating – Electrostatic Magic
W: www.electrostaticmagic.co.uk
Thread repair kits – Uni-Thread
T: 01803 867832.
Trailer manufacturers – Aluminium Trailer Company T: 01844 353539.
W: www.allytrailer.co.uk

NUTS, BOLTS & FIXINGS
LBF T: 01263 713498.
E: ray@lotusbendit.plus.com

PAINTING/BODYSHOP SERVICES
Auto Mirage T: 01253 734743.
W: www.automirage.co.uk
Beemabuild T: 0121 553 5550.
W: www.beemabuild.co.uk
Brooklands Motor Company
T: 01932 828545.
W: www.brooklandsmotorcompany.co.uk
IDL UK T: 01424 854900.
W: www.idluk.eu
Lee's Bodyshop T: 01332 331764.
W: www.leesautobodyshop.co.uk
Pinewood Body Repairs T: 01304 203020.
Precision Paint T: 01823 666289
W: www.precisionpaint.co.uk
Southside Accident And Repair Centre
T: 020 8317 1111.
W: www.southsidearc.com
SMS Autospray T: 01406 371504.
W: www.smsautospray.co.uk
Specialised Paintwork
T: 0118 930 6206.
W: www.specialisedpaintwork.com
The Colourworx T: 01637 873218.
W: www.thecolourworx.co.uk

PARTS SUPPLIERS (GENERAL BROCHURE)
Burton Power T: 020 8518 9189.
W: www.burtonpower.com
Cambridge Motorsport Parts
T: 01462 684300.
W: www.cambridgemotorsport.com
Car Builder Solutions T: 01580 891309.
W: www.cbsonline.co.uk
Demon Tweeks T: 0845 330 4751.
W: www.demon-tweeks.co.uk
Europa Spares T: 01283 815609.
W: www.europaspares.com
Furore T: 07905 897407.
W: www.forurecars.co.uk
Kit Parts Direct T: 07895 864500.
W: www.kitpartsdirect.com
Machine Mart T: 0844 8801250.
W: www.machinemart.co.uk
Merlin Motorsport T: 01249 782101.
W: www.merlinmotorsport.co.uk
Rally Design T: 01227 792792.
W: www.rallydesign.co.uk
Richbrook
W: www.richbrook-styling.co.uk

PROPSHAFT SERVICES
Autoprop T: 01342 322623.
W: www.autoprop-uk.co.uk
Bailey Morris T: 01480 216250.
W: www.baileymorris.com
CPS Drivelink T: 0191 4821690.
W: www.drivelink.com
Dunning & Fairbank
T: 0113 248 8788.
W: www.dandfltd.co.uk
Reco-Prop T: 01582 412110.
W: www.reco-prop.com

RUST PREVENTION
Electrostatic Magic
W: www.electrostaticmagic.co.uk
KBS Rustseal T: 01803 527961.
W: www.therustshop.com
Rustbuster T: 01775 761222.
W: www.rust.com

ROLLING ROAD/SUSPENSION TUNING
Atspeed T: 01268 773377.
W: www.atspeedracing.co.uk
Daytuner Performance
T: 01423 523323.
W: www.daytuner.co.uk
John Clarkson Autos
T: 01257 263879.
E: ajcmimi@tiscali.co.uk
Northampton Motorsport
T: 01604 766624.
W: www.northamptonmotorsport.com
Track Developments
T: 01666 840482.
W: www.trackdevelopments.co.uk

SEAT MANUFACTURERS/SUPPLIERS
Cobra Seats T: 01952 684020.
W: www.cobraseats.com
Corbeau Seats T: 01424 854499.
W: www.corbeau-seats.co.uk

Appendix: Useful Contacts

Intatrim T: 01952 608608.
W: www.intatrimtelford.co.uk
Interiors Seating T: 01623 400660.
W: www.interiorsseating.co.uk
JK Composites T: 01704 569730.
W: www.jkcomposites.com
Tillett Racing Seats T: 01795 420312.
W: www.tillett.co.uk

SUSPENSION COMPONENTS
Dampertech T: 01709 703992.
W: www.dampertech.co.uk
Protech Shocks T: 01225 705553.
W: www.protechshocks.co.uk
Superflex T: 01749 678152.
W: www.superflex.co.uk

TOOL SUPPLIERS
Draper T: 023 8049 4333.
W: www.draper.co.uk/ckc

Milli-Grip T: 01273 494844.
W: www.milli-grip.com
Memfast T: 01386 556868.
W: www.memfast.co.uk
Perm-Grit Tools T: 0800 298 5121.
W: www.permagrit.com

TRIM SERVICES
CC Trimming
T: 0121 558 9135.
W: www.cctrim.co.uk
Gabbat & Brown
T: 01704 821105.
W: www.gabbatandbrown.co.uk
M&M Classic Car Components
T: 01775 762004.
W: www.m-mclassiccartrim.com
Seals+Direct T: 0845 226 3345.
W: www.sealsplusdirect.co.uk
Woolies T: 01778 347347.
W: www.woolies-trim.co.uk

WHEEL SUPPLIERS
BK Racing W: www.bkracing.co.uk
Compomotive T: 01902 311499.
W: www.comp.co.uk
Force Racing T: 0113 252 5507.
W: www.force-racing.co.uk
Hawk Cars T: 01892 750282.
W: www.hawkcars.co.uk
Image Wheels T: 0121 522 2442.
W: www.imagewheels.co.uk
John Brown Wheels
W: www.johnbrownwheels.com
Momo T: 01268 764411.
W: www.momo-uk.co.uk
Performance Wheels T: 01530 517920.
W: www.performwheels.co.uk
Team Dynamics
W: www.team-dynamics.com
TSW T: 01908 625625.
W: www.tsw-wheels.co.uk
Wolfrace W: www.wolfrace.co.uk

CAR BUILDER SOLUTIONS

Accessories

Air Filters

Braking & Clutch

Cable/Wire

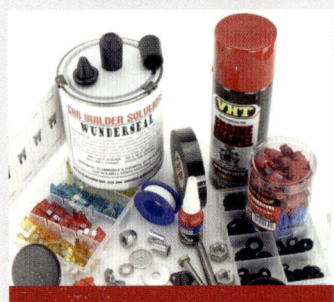
Consumables

Cooling System

Electrical

Exhaust System Parts

Fans

Fuel System

Coolant & Oil Tanks

Heating & Air-con

400 PAGE CATALOGUE FOR FREE
CALL OR ORDER ONE ONLINE

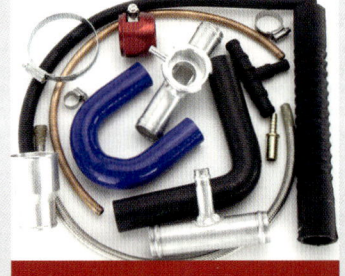
Hose Overbraid

Catches and Latches

www.carbuildersolutions.com

Redlands, Lindridge Lane, Staplehurst, Kent, TN12 0JJ

info@carbuildersolutions.com

Sales & Advice Tel: 01580 891309 or 01580 448007

Instruments

Insulation

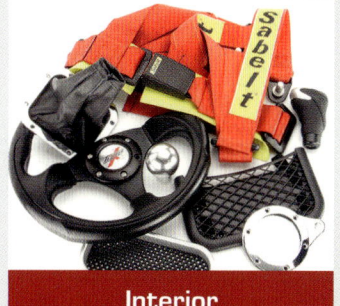
Interior

Lighting

Mechanical Parts

Mirrors

OVER 4500 PRODUCTS IN STOCK FOR IMMEDIATE SHIPPING

Nuts, Bolts & Fixings

P-Clips & Hose Clips

Sealant & Adhesive

Tools

Trim

Trimming & Upholstery

Tube, Sheet & Materials

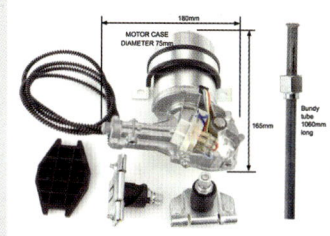
Windscreen Wash/Wipe

ALSO FROM complete KIT CAR magazine
BUILD · DRIVE · ENJOY

www.completekitcar.co.uk 01476 978843

UNTANGLE WIRING CONFUSION

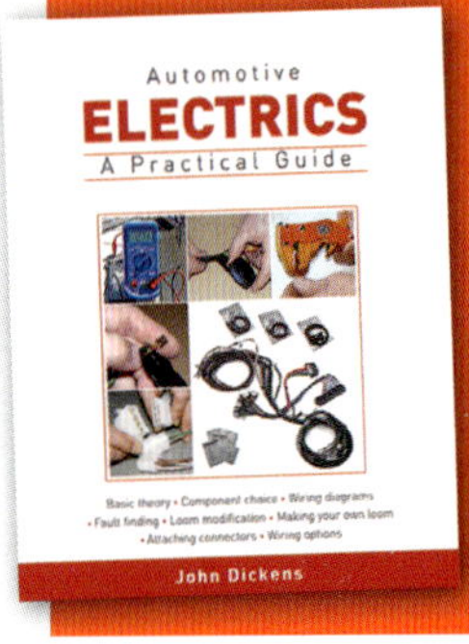

AUTOMOTIVE ELECTRICS BOOK
John Dickens

A practical guide that will give you an understanding of the principles of automotive wiring, from theory to making a loom from scratch, it concisely demystifies an aspect of kit car building that many fear.

£15 Plus P&P

LEARN DIY FIBREGLASS

AUTOMOTIVE FIBREGLASS BOOK
John Dickens

A clear and concise guide to fibreglass repair and fabrication. Step-by-step hands-on guides to a number of basic and complex projects makes this a really useful guide to working with the material.

£15 Plus P&P

HOW TO BE A KIT CAR MANUFACTURER

DESIGN AND BUILD A SPORTS CAR BOOK
Stuart Mills

Prolific kit car desginer and MEV founder Stuart Mills takes you through the fascinating process of designing and building a sports car, either as a one-off or with a view to going onto production. Stop dreaming and start building!

£15 Plus P&P